T0178576

SpringerBriefs in Molecular Science

Ultrasound and Sonochemistry

Series editors

Bruno G. Pollet, Faculty of Engineering, Norwegian University of Science
and Technology, Trondheim, Norway
Muthupandian Ashokkumar, School of Chemistry, University of Melbourne,
Melbourne, VIC, Australia

SpringerBriefs in Molecular Science: Ultrasound and Sonochemistry is a series of concise briefs that present those interested in this broad and multidisciplinary field with the most recent advances in a broad array of topics. Each volume compiles information that has thus far been scattered in many different sources into a single, concise title, making each edition a useful reference for industry professionals, researchers, and graduate students, especially those starting in a new topic of research.

More information about this series at http://www.springer.com/series/15634

Kyuichi Yasui

Acoustic Cavitation
and Bubble Dynamics

Kyuichi Yasui
National Institute of Advanced Industrial
  Science and Technology (AIST)
Moriyama-ku, Nagoya
Japan

ISSN 2191-5407   ISSN 2191-5415   (electronic)
SpringerBriefs in Molecular Science
ISSN 2511-123X   ISSN 2511-1248   (electronic)
Ultrasound and Sonochemistry
ISBN 978-3-319-68236-5   ISBN 978-3-319-68237-2   (eBook)
https://doi.org/10.1007/978-3-319-68237-2

Library of Congress Control Number: 2017952916

Printed on acid-free paper

This Springer imprint is published by Springer Nature
The registered company is Springer International Publishing AG
The registered company address is: Gewerbestrasse 11, 6330 Cham, Switzerland

Preface

Acoustic cavitation is the formation and subsequent collapse of bubbles in liquid irradiated with a powerful ultrasonic wave. Bubble dynamics are dynamics of bubble pulsation under intense ultrasound. Under certain conditions, a bubble violently collapses, resulting in high temperature and pressure inside a bubble. Light is emitted from a heated bubble (sonoluminescence), and chemical reactions take place inside the bubble (sonochemical reactions). Acoustic cavitation is useful for ultrasonic cleaning and sonochemistry. Many researchers have studied its medical applications such as cancer treatment and extracorporeal shock wave lithotripsy. Although there is no description on the medical applications herein, the description in this book on fundamental phenomena should be useful for readers who will study medical applications.

Although the phenomena have been studied for more than 100 years, considerable development in this field was brought about after the re-discovery of single-bubble sonoluminescence by Gaitan and Crum in 1989 (there is also an experimental report on single-bubble sonoluminescence published in 1962 [Young FR (2005) Sonoluminescence. CRC Press, Boca Raton]). Experimental evidence of plasma formation inside a bubble was found in optical spectra of single-bubble sonoluminescence in sulfuric acid by Flannigan and Suslick in 2005.

The present SpringerBrief in *Ultrasound and Sonochemistry* is written as an introduction to this field for students, researchers, engineers, educators, and teachers. For this purpose, many illustrations are added in order to help readers to understand the phenomena at a glance. Detailed derivation of mathematical equations of bubble dynamics is described for readers who will study the phenomena theoretically and numerically. There is no problem to skip such mathematical descriptions for readers who just want to understand the phenomena qualitatively. Chapter 1 focuses on acoustic cavitation, which is an introduction to the phenomena with many illustrations and photographs. Chapter 2 describes bubble dynamics, which most benefits readers who will study the phenomena theoretically and numerically. Chapter 3 highlights unsolved problems, which is written mostly for students and researchers who will work in this field.

The author would like to thank Profs. Bruno G. Pollet and Muthupandian Ashokkumar who recommended him to write this book and reviewed it. The author also would like to thank his collaborators in his research: Toru Tuziuti, Yasuo Iida, Wataru Kanematsu, Kazumi Kato, Noriya Izu, Atsuya Towata, Hideto Mitome, Nobuhiro Aya, Teruyuki Kozuka, Shin-ichi Hatanaka, Judy Lee, Sivakumar Manickam, Muthupandian Ashokkumar, Franz Grieser, and others. Finally, the author would like to thank the staff at Springer.

Nagoya, Japan   Kyuichi Yasui
August 2017

Contents

1 Acoustic Cavitation 1
   1.1   What Is Acoustic Cavitation? 1
   1.2   Power Ultrasound and Diagnostic Ultrasound 3
   1.3   Ultrasonic Transducers 4
   1.4   Ultrasonic Horn and Bath 6
   1.5   Traveling and Standing Waves 9
   1.6   Transient and Stable Cavitation 14
   1.7   Vaporous and Gaseous Cavitation 16
   1.8   Bubble Structures 17
   1.9   Sonoluminescence 19
   1.10 Sonochemistry 26
   1.11 Ultrasonic Cleaning 29
   References 31

2 Bubble Dynamics 37
   2.1   Rayleigh–Plesset Equation 37
   2.2   Rayleigh Collapse 41
   2.3   Keller Equation 42
   2.4   Method of Numerical Simulations 47
   2.5   Non-equilibrium Evaporation and Condensation 51
   2.6   Liquid Temperature at the Bubble Wall 53
   2.7   Gas Diffusion (Rectified Diffusion) 54
   2.8   Chemical Kinetic Model 56
   2.9   Single-Bubble Sonochemistry 56
   2.10 Main Oxidants 61
   2.11 Effect of Volatile Solutes 64
   2.12 Resonance Radius 68
   2.13 Shock Wave Emission 73
   2.14 Shock Formation Inside a Bubble 75
   2.15 Jet Penetration Inside a Bubble 76

2.16   Radiation Forces (Bjerknes Forces) 78
2.17   Effect of Salts and Surfactants 83
2.18   Bubble–Bubble Interaction 84
2.19   Acoustic Cavitation Noise 86
2.20   Acoustic Streaming and Microstreaming 92
References 93

3   Unsolved Problems 99
3.1   Cavitation Nuclei (Bulk Nanobubbles) 99
3.2   Ammonia (NH_3) Formation 106
3.3   Solidification and Sonocrystallization 107
3.4   A Hot Plasma Core 108
3.5   Ionization-Potential Lowering 110
3.6   OH-Line Emission 112
3.7   Acoustic Field 116
3.8   Effect of a Magnetic Field 116
3.9   Role of Oxygen Atoms 117
3.10   Extreme Conditions in a Dissolving Bubble 117
3.11   Concluding Remarks (Modeling Complex Phenomena) 118
References 118

Chapter 1
Acoustic Cavitation

Abstract Acoustic cavitation is the formation and subsequent violent collapse of bubbles in liquid irradiated with intense ultrasound. Ultrasound is radiated by a vibrating plate connected to ultrasonic transducers made of piezoelectric materials driven by electrical power. Microscopic mechanism for vibration of piezoelectric materials is briefly described. There are two types of ultrasonic experimental equipment used to generate acoustic cavitation: ultrasonic horn (or probe) and ultrasonic bath. Ultrasonic standing waves and traveling waves are discussed by means of mathematical equations. Acoustic impedance is discussed, and transmission and reflection coefficients are described. Various types of acoustic cavitations are discussed: transient and stable cavitations, vaporous and gaseous cavitations. Fluctuations in degassing and re-gassing cause repeated change between vaporous and gaseous cavitation. Light emission associated with violent bubble collapse as well as chemical reactions inside and outside a bubble is discussed in the sections entitled "sonoluminescence" and "sonochemistry," respectively. Unsolved problems in sonoluminescence are briefly discussed. Reasons for lesser amount of produced H radicals (H·) than that of OH radicals (OH·) in sonochemical reactions are discussed based on results generated from numerical simulations. In the last section, ultrasonic cleaning, especially for the application to silicon wafers, is discussed.

Keywords Negative pressure · Bolt-clamped Langevin-type transducer Resonance · Acoustic impedance · Damped standing wave · Cavitation oscillation Acoustic Lichtenberg figure · Plasma formation · Reactions of OH radicals Megasonic

1.1 What Is Acoustic Cavitation?

An acoustic wave (sound) is a propagation of pressure oscillation with sound velocity in a medium such as liquid, gas, or solid (Fig. 1.1) [1–3]. Ultrasound is an inaudible sound with a frequency of pressure oscillation higher than 20 kHz

© The Author(s) 2018   1
K. Yasui, *Acoustic Cavitation and Bubble Dynamics*,
Ultrasound and Sonochemistry, https://doi.org/10.1007/978-3-319-68237-2_1

(= 2×10^4 cycles/s). Most of the time, ultrasound is defined as an acoustic wave with a frequency higher than 10 kHz for convenience. It should be noted that some young people can often hear ultrasound at around 20 kHz. In addition, due to some nonlinear effects, audible sound with a frequency less than 20 kHz is sometimes radiated into atmosphere in ultrasonic experiments. Thus, in experiments using intense ultrasound at relatively low frequencies, earplugs or headphones should be used due to the possible health risks.

The wavelength (λ) of an acoustic wave is defined as the length for one pressure oscillation (Fig. 1.1). The acoustic period (T_a) is defined as time for one pressure oscillation. The frequency (f) of an acoustic wave is defined as the number of pressure oscillation per unit time (second): $f = 1/T_a$. The sound velocity (or sound speed) (c) is defined as the distance for a pressure disturbance propagating per unit time: $c = f\lambda$. The sound velocity in dry air and liquid water at room temperature is about 340 m/s and 1500 m/s, respectively. The sound velocity in liquid water increases with temperature and has a maximum value of about 1555 m/s at around 74 °C. The acoustic pressure amplitude (p_a) is defined as the amplitude for pressure oscillation (Fig. 1.1).

When liquid such as water is irradiated under intense ultrasound, many tiny gas bubbles appear. In the rarefaction phase of the ultrasonic wave, instantaneous local pressures in liquid become *negative* when the acoustic pressure amplitude is larger than the ambient pressure (normally, the ambient pressure is $p_\infty = 1$ atm $= 1.01325$ bar $= 1.01325 \times 10^5$ Pa, where 1 bar $= 10^5$ Pa and 1 Pa $= 1$ N/m^2). *Negative* pressure is possible only in liquids or solids, and impossible in gases [4]. This is the "force" to expand a liquid (or solid) element (Fig. 1.2) [5]. As a result, gases dissolved in the liquid appear as gas bubbles because gases can no longer be dissolved in the liquid under negative pressures. During the ultrasonic wave

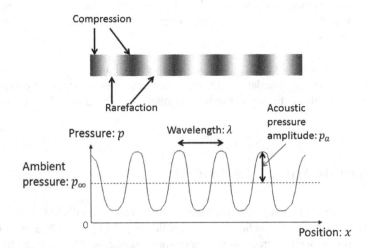

Fig. 1.1 Acoustic wave (ultrasound)

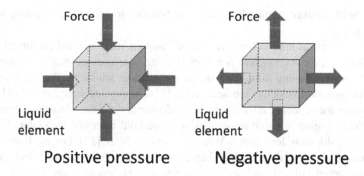

Fig. 1.2 Negative pressure. Reprinted with permission from Yasui et al. [5]. Copyright (2004), Taylor and Francis

rarefaction phase, many tiny bubbles expand as the pressure at the bubble wall is higher than the liquid pressure at a distance from the bubble. During the compression phase of the ultrasonic wave, some of the bubbles violently collapse leading to shock wave being emitted into the liquid [6]. Near a solid surface, a liquid jet penetrates into the bubble toward it, in turns causing surface erosion. The phenomenon of the bubble formation and subsequent violent collapse of the bubble under an acoustic wave (ultrasound) is called *acoustic cavitation.*

There are chiefly two points which make acoustic cavitation different from boiling. One is that the reduction in pressure is the origin for the bubble formation in acoustic cavitation, while heating is the origin in boiling. The other is the presence and absence of the violent bubble collapse in acoustic cavitation and boiling, respectively.

1.2   Power Ultrasound and Diagnostic Ultrasound

Power ultrasound often refers to ultrasound with its intensity higher than the threshold intensity for violent bubble collapse. The threshold for violent bubble collapse is often different from that for acoustic cavitation (bubble nucleation) to occur because the latter strongly depends on degree of gas saturation in liquid (see Fig. 3.1) [7]. Typical threshold pressure amplitude for violent bubble collapse increases as ultrasonic frequency increases: about 1.2 atm at 20 kHz, 1.6 atm at 140 kHz, 3 atm at 1 MHz (= 10^6 Hz), and 5.8 atm at 5 MHz [8, 9]. For a plane or a spherical traveling wave of ultrasound in liquid water, they correspond to the following intensity of ultrasound which is defined as the average rate of flow of energy through a unit area normal to the direction of ultrasound propagation: 0.49 W/cm^2 at 20 kHz, 0.88 W/cm^2 at 140 kHz, 3 W/cm^2 at 1 MHz, and 11 W/cm^2 at 5 MHz (see Eq. (1.1) in the next section) [10]. It should be noted that the threshold intensity may be different for a standing wave of ultrasound, although the threshold pressure amplitude is the same between traveling and standing waves (see Sect. 1.5).

Diagnostic ultrasound often refers to ultrasound used in medical imaging of fetus, abdomen, etc. Ultrasound is reflected at the interfaces of internal organs and biological tissues having different acoustic impedances which are the mass density multiplied by sound velocity (see Sect. 1.5) [11]. The medical imaging is conducted by detecting the reflected ultrasound. For safety criterion, the ultrasound intensity for medical imaging is much lower than the threshold intensity for violent bubble collapse, considerably less than 1 W/cm^2 at several MHz [12]. For applications of ultrasound to therapy, on the other hand, acoustic cavitation with violent bubble collapse is often utilized such as in cancer treatment, extracorporeal shock wave lithotripsy, using ultrasound in MHz range with much higher intensity [13].

1.3 Ultrasonic Transducers

A vibrating plate radiates an acoustic wave with its frequency identical to the frequency of vibration [10]. Piezoelectric materials such as crystallized quarts (SiO$_2$), barium titanate (BaTiO$_3$), PZT (lead zirconate titanate) (Pb(Zr$_x$, Ti$_{1-x}$)O$_3$) vibrate with the frequency of an AC voltage applied to the material. In other words, the frequency of ultrasound radiated from an ultrasonic transducer is the same as that of an AC voltage applied to a transducer.

The piezoelectric effect is the formation of electric dipoles by the application of pressure (stress) on a material [14]. Inverse piezoelectric effect is the deformation of a material by the application of an electric field. A schematic representation of the mechanism of inverse piezoelectric effect is shown in Fig. 1.3. By the application of

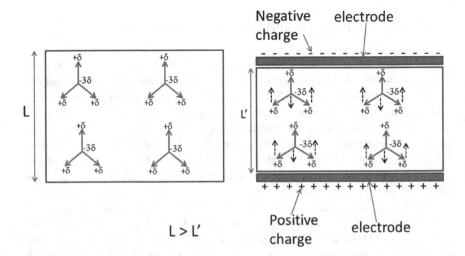

Fig. 1.3 Mechanism of inverse piezoelectric effect

an electric field, positively charged ions in a crystal of a piezoelectric material slightly move toward the negatively charged electrode. On the contrary, negatively charged ions slightly move toward the positively charged electrode. As a result, piezoelectric material is deformed by the application of an electric field.

A piece of piezoelectric material vibrates most strongly when it is driven at its resonance frequency (f_0). Resonance frequency is determined by the mass and stiffness of a piezoelectric material. In other words, the resonance frequency is determined by the volume and shape of a material if the density of a material is kept constant. Generally speaking, the resonance frequency decreases as the volume of a material increases. It should be noted, however, there are multiple resonance frequencies for a piece of material. In a simple case, an integer multiple of the fundamental resonance frequency is also a resonance frequency, which is called a *harmonic frequency* (or higher order resonance frequency).

For ultrasonic irradiation, a thin plate of a piezoelectric material is used in combination with a vibration plate for high frequencies in the range of 100 kHz– 1 MHz (= 10^6 Hz) (Fig. 1.4) [15]. For low ultrasonic frequencies (20–200 kHz), bolt-clamped Langevin-type transducers (BLT) are used (Fig. 1.4). The BLT was invented by Paul Langevin (1872–1946) who was a French physicist. In BLT, the piezoelectric ceramic is tightly sandwiched between two pieces of metal with a bolt. This compression of piezoelectric ceramic materials enables high amplitude oscillations of piezoelectric ceramic of low tensile strength. Moreover, resonance frequency is considerably decreased by the presence of metal blocks, and the BLT is more suited for low ultrasonic frequencies.

To estimate the acoustic pressure amplitude (p_a), the acoustic intensity (I) is often used because the acoustic intensity is related to the acoustic pressure amplitude (for a plane or a spherical sinusoidal wave) [10]. Here, the acoustic

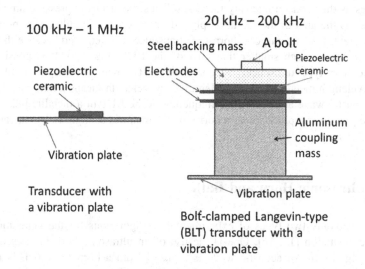

Fig. 1.4 Typical ultrasonic transducers with a vibration plate

Fig. 1.5 A horn-type
transducer

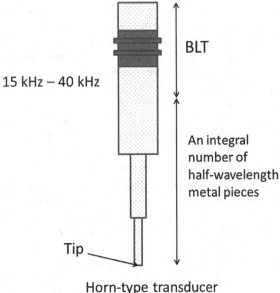

BLT

15 kHz – 40 kHz

An integral
number of
half-wavelength
metal pieces

Tip

Horn-type transducer

intensity (*I*) is defined as the average rate of flow of energy through a unit area normal to the direction of propagation. The units are in W/m^2.

$$I = \frac{p_a^2}{2\rho_0 c} \tag{1.1}$$

where ρ_0 is the density of a medium (liquid). The acoustic pressure amplitude increases as the acoustic intensity increases. Thus, the acoustic pressure amplitude increases as the area of the vibration plate decreases when the ultrasonic power is kept constant. For this purpose, horn-type transducer is often used because the area of a horn tip is much smaller than that of the BLT (Fig. 1.5). It is possible to fabricate horn-type transducer by connecting BLT with an integer number of half-wavelength metal pieces [15]. Here, half wavelength means a half wavelength of an acoustic wave at a resonance frequency of the BLT in a metallic piece. The acoustic pressure amplitude near a horn tip is sometimes as high as 10 bar (around 10 atm) or more.

1.4   Ultrasonic Horn and Bath

There are mainly two types of experimental configurations for the generation of acoustic cavitation [15, 16]. One is the use of an ultrasonic horn immersed in a liquid (Fig. 1.6a). An acoustic wave is radiated from a horn tip which is much smaller than the acoustic wavelength. The other is the use of an ultrasonic bath

Fig. 1.6 Three experimental methods for the generation of acoustic cavitation. **a** Horn type, **b** bath type, and **c** indirect irradiation in bath type. Reprinted with permission from Yasui et al. [16]. Copyright (2005), Elsevier

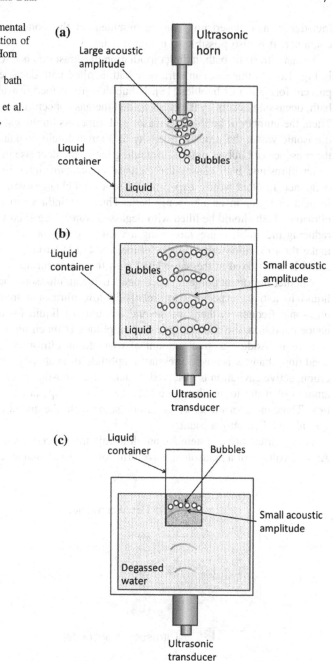

usually used in ultrasonic cleaning (Fig. 1.6b). One of the several ultrasonic transducers is attached to outer surface of a liquid container or attached to inner surface of a small closed box immersed into the liquid [17]. An immersible transducer in a closed box can be mounted on the bottom or wall of a liquid container. It is also possible to mount it using a rack.

For an ultrasonic bath, indirect irradiation of ultrasound is also possible as shown in Fig. 1.6c. In this case, an ultrasonic bath is filled with degassed water in order to prevent formation of bubbles [18]. If bubbles are formed in water in an ultrasonic bath, degassing occurs by the bubbles, and the gas concentration in water decreases. Then, the number of bubbles decreases with time. As bubbles strongly attenuate the ultrasonic wave, the acoustic intensity in a small liquid container is influenced by the presence of bubbles in the surrounding bath. The decrease in number of bubbles in an ultrasonic bath causes the increase in acoustic intensity in a small liquid container. In other words, experimental condition changes with time if bubbles are formed in water in an ultrasonic bath. Thus, for indirect irradiation method, an ultrasonic bath should be filled with degassed water. Degassed water is prepared by reducing the ambient pressure using a vacuum pump or by boiling it. In order to make the irradiation condition the same, the liquid surface in a small liquid container is often fixed at the same level with the water surface in a bath.

Another important point in experiments using an ultrasonic bath is the amount of liquid in a bath, especially for relatively low ultrasonic frequencies. Under an ultrasonic frequency, there are several amounts of liquid for resonance. At resonance condition, the acoustic pressure amplitude is much higher than that in other conditions. Adding a small amount of water to an ultrasonic bath near resonance condition changes acoustic pressure amplitude dramatically. At a resonance condition, active cavitation is observed visually and acoustically. By adding some extra amount of water to an ultrasonic bath, cavitation stops at an antiresonance condition. Thus, in experiments using an ultrasonic bath, the quantity of liquid in a bath should be taken into account.

An electrical drive system for an ultrasonic transducer is shown in Fig. 1.7 [19]. An AC voltage of a sinusoidal waveform for a desired frequency is generated by a

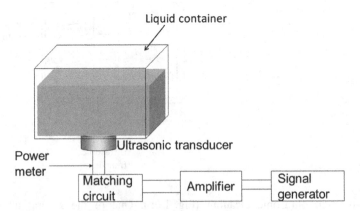

Fig. 1.7 An electrical drive system for an ultrasonic transducer. Reprinted with permission from Yasui [19]. Copyright (2011), Springer

signal generator. The AC power is amplified by a power amplifier. In order to prevent a reflection of the AC power at an ultrasonic transducer, an appropriate matching circuit is inserted between an ultrasonic transducer and a power amplifier. The electric power used in an ultrasonic transducer is measured by a power meter inserted between an ultrasonic transducer and a matching circuit. The electrical drive system is essentially the same for an ultrasonic horn.

1.5 Traveling and Standing Waves

For a plane traveling (or progressive) wave of sound, the acoustic pressure (p) is given as follows [10].

$$p = p_a \sin(kx - \omega t) \tag{1.2}$$

where k is the wave number, x is the position in the direction of the wave propagation, ω is the angular frequency of sound ($\omega = 2\pi f$), and t is time. The wave number is related to the angular frequency and sound velocity as $k = \omega/c$. It is straightforward to show that the velocity of propagation of an acoustic wave described by Eq. (1.2) is actually equivalent to the sound velocity (c). For example, a fixed phase angle in sinusoidal function in Eq. (1.2) is given as follows.

$$(kx - \omega t) = \text{const.} \tag{1.3}$$

The velocity of propagation of a fixed phase is derived by differentiating both sides of Eq. (1.3) with time (t) as follows.

$$k\frac{dx}{dt} - \omega = 0 \tag{1.4}$$

thus,

$$\frac{dx}{dt} = \frac{\omega}{k} = c \tag{1.5}$$

In acoustic wave propagation in a medium, a small element (volume) of medium moves forward and backward around an equilibrium position [20]. Moreover, a small element which is often called "a particle" expands and contracts according to the pressure oscillation of sound. The velocity of a small element is called particle velocity (u) and is given as follows [10].

$$u = \frac{p_a}{\rho_0 c} \sin(kx - \omega t) \tag{1.6}$$

where ρ_0 is equilibrium density of a medium (liquid). Acoustic impedance (z) is defined as the ratio of acoustic pressure to particle velocity [10].

$$z = \frac{p}{u} \tag{1.7}$$

For a plane traveling wave propagating in a positive x direction, $z = \rho_0 c$ from Eqs. (1.2) and (1.6). At 20 °C and 1 atmosphere (1 atm), the sound velocity and density of distilled water are 1482 m/s and 998 kg/m^3, respectively. Thus,

$$(\rho_0 c)_{20 \, °C} = 1.48 \times 10^6 \quad Pa \cdot s/m \tag{1.8}$$

If the acoustic pressure amplitude is 1 atm (= 1.01325×10^5 Pa), then the amplitude of the particle velocity is $\frac{p_a}{\rho_0 c} = 0.0676$ m/s $= 6.76$ cm/s. It should be noted that the particle velocity is different from velocity of the liquid flow (*acoustic streaming*) [21].

For a plane traveling wave propagating in a *negative* x direction, the acoustic pressure and particle velocity are expressed as follows [10].

$$p = p_a \sin(kx + \omega t) \tag{1.9}$$

$$u = -\frac{p_a}{\rho_0 c} \sin(kx + \omega t) \tag{1.10}$$

Thus, acoustic impedance in this case is $z = -\rho_0 c$.

When a point source radiates an acoustic wave into a medium (liquid), a spherical wave is formed as Eq. (1.11).

$$p = \frac{p_a}{r} \sin(kr - \omega t) \tag{1.11}$$

where r is distance from a point source. In this case, particle velocity is given as follows.

$$u = \frac{1}{\rho_0 c} \frac{p_a}{r} \frac{1}{\cos \theta} \sin(kr - \omega t + \theta) \tag{1.12}$$

where θ is given by $\tan \theta = 1/kr$. In other words, the particle velocity is not in phase with the pressure in contrast to plane waves. The derivation of Eq. (1.12) is given in Ref. [10].

When a plane circular piston radiates an acoustic wave into a medium (liquid), the spatial distribution of the acoustic pressure amplitude is described by Eq. (1.13) on the symmetry axis [10].

$$p_a(x) = 2\rho_0 c v_0 \left| \sin\left(\frac{\pi}{\lambda}\left(\sqrt{x^2 + a^2} - x\right)\right) \right| \qquad (1.13)$$

where v_0 is the velocity amplitude of a vibrating circular piston, λ is the wavelength of an acoustic wave in a medium (liquid), x is the distance from a circular piston on the symmetry axis, and a is the radius of a circular piston. A tip of an ultrasonic horn is similar to a circular piston. When the bubbles are formed by ultrasound, however, the density and sound velocity in a medium as well as the wavelength in Eq. (1.13) change by the presence of bubbles. Generally speaking, the density and sound velocity decrease by the presence of bubbles. Thus, the acoustic pressure amplitude drops by the presence of bubbles under an ultrasonic horn. In fact, the drop in acoustic pressure amplitude has been experimentally observed under an ultrasonic horn [22].

In an ultrasonic bath, the acoustic wave is reflected both at the liquid surface and inner walls of the bath. Especially, reflection at the liquid surface is almost complete as discussed below. The pressure reflection and transmission coefficients are defined as follows [10].

$$R = \frac{p_{a,r}}{p_{a,i}} \qquad (1.14)$$

$$T = \frac{p_{a,t}}{p_{a,i}} \qquad (1.15)$$

where R and T are the pressure reflection and transmission coefficients, respectively, $p_{a,i}$ is the acoustic pressure amplitude of an incident wave, $p_{a,r}$ and $p_{a,t}$ are the acoustic pressure amplitudes of reflected and transmitted waves, respectively. $p_{a,r}$ takes a negative value when there is a phase shift of π at the reflection. The intensity reflection and transmission coefficients are defined as follows.

$$R_I = \frac{I_r}{I_i} = |R|^2 \qquad (1.16)$$

$$T_I = \frac{I_t}{I_i} = \frac{r_2}{r_1} |T|^2 \qquad (1.17)$$

where R_I and T_I are intensity reflection and transmission coefficients, respectively, I_i is the acoustic intensity of an incident wave, I_r and I_t are the acoustic intensities of the reflected and transmitted waves, respectively, r_1 and r_2 are the characteristic acoustic impedances of the medium 1 and 2 defined as $r_1 = \rho_1 c_1$ and $r_2 = \rho_2 c_2$, respectively, medium 1 and 2 are media for the initial incident wave and for the transmitted wave, respectively, and ρ_1 and c_1 (ρ_2 and c_2) are the equilibrium density and sound velocity of medium 1 (2), respectively. Equations (1.16) and (1.17) are derived from the fact that the acoustic intensity is given by $p_{a,i}^2/2r_i$ for a plane traveling wave according to Eq. (1.1), where subscript i indicates medium 1 or 2.

When a plane traveling wave is reflected at normal incidence on a planar interface between two media, the pressure reflection and transmission coefficients can be derived [10]. The incident and transmitted waves propagate in a positive x direction, and a reflected wave propagates in a negative x direction. The boundary conditions at a planar interface (at $x = 0$) are expressed as follows.

$$p_{a,i} + p_{a,r} = p_{a,t} \tag{1.18}$$

$$u_{a,i} + u_{a,r} = u_{a,t} \tag{1.19}$$

where $u_{a,i}$ is the amplitude of the particle velocity for an incident wave, $u_{a,r}$ and $u_{a,t}$ are the amplitudes of the particle velocity for the reflected and transmitted waves, respectively, and each amplitude of particle velocity can possibly take a negative value. The division of Eq. (1.18) by Eq. (1.19) yields

$$\frac{p_{a,i} + p_{a,r}}{u_{a,i} + u_{a,r}} = \frac{p_{a,t}}{u_{a,t}} \tag{1.20}$$

Since a plane traveling wave has $z = \frac{p}{u} = \pm \rho_i c_i = \pm r_i$, depending on the direction of propagation, Eq. (1.20) becomes

$$r_1 \frac{p_{a,i} + p_{a,r}}{p_{a,i} - p_{a,r}} = r_2 \tag{1.21}$$

which yields the following pressure reflection coefficient defined in Eq. (1.14).

$$R = \frac{1 - r_1/r_2}{1 + r_1/r_2} \tag{1.22}$$

From Eq. (1.18), $1 + R = T$ holds.

$$T = \frac{2}{1 + r_1/r_2} \tag{1.23}$$

The negative value of R means that a phase (a phase angle in a sinusoidal function) of the reflected wave is shifted by π. The intensity reflection and transmission coefficients are accordingly given as follows.

$$R_I = \left(\frac{1 - r_1/r_2}{1 + r_1/r_2} \right)^2 \tag{1.24}$$

$$T_I = 4 \frac{r_1/r_2}{(1 + r_1/r_2)^2} \tag{1.25}$$

Thus, the intensity reflection coefficient is nearly 1 when the characteristic acoustic impedances of medium 1 and 2 are largely different ($r_1 \gg r_2$ or $r_1 \ll r_2$). In other words, at the interface between a liquid and a gas, an ultrasonic wave is completely reflected. For example, at 20 °C, 1 atm, with a density of air of 1.21 kg/m^3 and a sound velocity of 343 m/s, an acoustic impedance of 415 Pa s/m is found. As the characteristic acoustic impedance of water at 20 °C is 1.48×10^6 Pa s/m, the intensity reflection coefficient is 0.9989 for both cases of an incident wave propagating in water and air (the intensity transmission coefficient is 1.1×10^{-3}).

In an ultrasonic bath, an ultrasonic wave is almost completely reflected at the liquid surface. When an ultrasonic transducer is attached to a side wall of a bath, an ultrasonic wave is reflected at the other side of the bath wall. As a result, a standing wave is formed. When an ultrasonic transducer is attached to the bottom of an ultrasonic bath, a positive x direction is taken toward a planar liquid surface from the bottom of a bath. The planar liquid surface is at $x = 0$ (the bottom of the bath is at $x = -L < 0$, where L is the liquid height). Then, an ultrasonic wave radiated from an ultrasonic transducer at the bottom and a reflected wave from a liquid surface are expressed as follows [10].

$$p_i = p_{a,i} \sin(-kx + \omega t) \tag{1.26}$$

$$p_r = -p_{a,i} \sin(kx + \omega t) \tag{1.27}$$

where p_i and p_r are the acoustic pressures of an incident and reflected waves, respectively, and the pressure reflection coefficient in Eq. (1.22) is approximated as $-1(-0.9994)$. From the principle of superposition for linear ultrasonic waves, the acoustic pressure in liquid is expressed as follows.

$$p = p_i + p_r = p_{a,i}[\sin(-kx + \omega t) - \sin(kx + \omega t)] = 2p_{a,i} \cos(\omega t) \sin(-kx) \tag{1.28}$$

where the formula for sinusoidal functions $\left(\sin A - \sin B = 2 \cos \frac{A+B}{2} \sin \frac{A-B}{2}\right)$ is used. Equation (1.28) is a surprising result because the waveform $\left(2p_{a,i} \sin(-kx)\right)$ does not change with time and temporal oscillation of acoustic pressure is only in phase $\left(2p_{a,i} \sin(-kx) > 0\right)$ or in antiphase $\left(2p_{a,i} \sin(-kx) < 0\right)$. For several planes including the liquid surface ($x = 0$) which satisfy $x = -\frac{n\pi}{k} = -\frac{n}{2}\lambda$, where n is an integer and λ is the ultrasonic wavelength in liquid, the acoustic pressure is always zero. These planes are called *pressure nodes*. On the contrary, for several other planes satisfying $x = -\frac{\left(n\pi - \frac{\pi}{2}\right)}{k} = \left(-\frac{n}{2} + \frac{1}{4}\right)\lambda$, absolute value of acoustic pressure amplitude takes maximum value. This type of planes is called *pressure antinodes*. The distance between successive nodes (antinodes) is a half wavelength ($\lambda/2$). The distance between neighboring node and antinode is a quarter wavelength ($\lambda/4$). The liquid surface is always a pressure node.

Fig. 1.8 A damped standing wave. Reprinted with permission from Yasui [23]. Copyright (2016), Springer

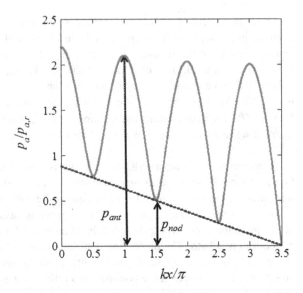

In actual experiments using an ultrasonic bath, however, there is considerable attenuation of ultrasound in a bubbly liquid. In this case, an acoustic wave is a mixture of traveling and standing waves as shown in Fig. 1.8, which is often called a *damped standing wave* [10, 23]. At pressure "nodes" (planes for local minimum in acoustic pressure amplitude), acoustic pressure amplitude is no longer zero except at the liquid surface (the right side in Fig. 1.8). The straight dotted line in Fig. 1.8 shows the traveling wave component. The percentage of a standing wave component is sometimes defined as follows [1].

$$\frac{(p_{ant} - p_{nod})}{(p_{ant} + p_{nod})} \times 100\% \tag{1.29}$$

where p_{ant} and p_{nod} are the acoustic pressure amplitudes at a pressure "antinode" and "node," respectively, as shown in Fig. 1.8. The value defined in Eq. (1.28) depends upon the distance from the bottom of a liquid container where an ultrasonic transducer is attached ($x = 0$ in Fig. 1.8).

1.6 Transient and Stable Cavitation

There are mainly two types of acoustic cavitation. One is *transient cavitation* and the other is *stable cavitation* [19]. There are two varying definitions of transient cavitation. One is that lifetime of a bubble is only one or a few acoustic cycles

Fig. 1.9 A photograph of single-bubble sonoluminescence (a point at the center of a liquid container). Courtesy of Dr. S. Hatanaka

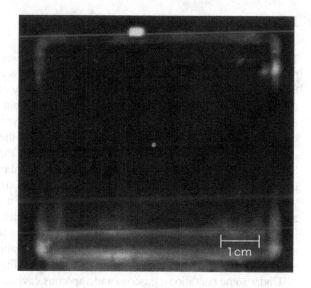

because a bubble is fragmented into "daughter" bubbles due to its shape instability. The other is that a bubble undergoes strong collapse resulting in light emission (sonoluminescence) and/or chemical reactions (sonochemical reactions). Accordingly, stable cavitation is defined in two different ways. One is that the lifetime of a bubble is very long. The other is that a bubble pulsates mildly without any light emission and chemical reactions.

There are some bubbles which lead to both transient and stable cavitation depending upon the definition. In other words, some bubbles are active in light emission and chemical reactions but have a long lifetime. This type of bubbles is known in single-bubble sonoluminescence (SBSL) [24] (Fig. 1.9). In SBSL experiments, a single bubble is trapped near the pressure antinode of an ultrasonic standing wave (mostly at low frequencies such as 20–50 kHz). A SBSL bubble stably repeats expansion and contraction for a long period of time (for even several days!). At each bubble collapse, a faint light is emitted from a SBSL bubble. The light pulse is emitted repeatedly every acoustic cycle like a clock. In a dark room, light of SBSL is visible to the naked eyes like a star in the sky because for human eyes, SBSL light pulses are seen as a continuous light from a point due to the very high repetition frequency of light emissions. Mechanism(s) of light emission is discussed in Sect. 1.8. Such a bubble is classified as a stable cavitation bubble according to the first definition and called "high-energy stable cavitation" bubbles [1]. From the second definition, however, it is classified into a transient cavitation bubble and named "repetitive transient cavitation" bubbles. When the terms transient and stable cavitation are used, it is necessary to indicate which definition is used: lifetime (shape stability) or activity.

1.7 Vaporous and Gaseous Cavitation

There is another classification of cavitation: *vaporous* and *gaseous cavitation* [25, 26]. There are mainly two definitions for vaporous cavitation. One is that the bubble content is mostly (water) vapor. This is widely observed at low ultrasonic frequencies with relatively high acoustic pressure amplitude because the evaporation of (water) vapor during the bubble expansion is very intense due to the very large expansion of a bubble. The other definition is the cavitation in a partially degassed (undersaturated) liquid (water). In this case, there are only a few visible bubbles in the liquid, and mist formation from a "fountain" at the liquid surface is intense (Fig. 1.10) [26]. Accordingly, gaseous cavitation is defined in two different ways. One is that the bubble content is mostly made of noncondensable gas such as air. This is widely observed at high ultrasonic frequencies. The other is cavitation in liquid (water) nearly saturated or oversaturated with gas (air). In this case, there are many visible gas bubbles in the liquid, and mist formation at the liquid surface is less intense compared to that in vaporous cavitation (Fig. 1.10) [26].

Under some conditions, gaseous and vaporous cavitations defined by the second definition occur alternately in a timescale of 100 s without changes in experimental conditions (Fig. 1.10) [26]. It is called *cavitation oscillation*. The reason for cavitation oscillation is the repeated degassing and re-gassing (dissolving of gas into liquid). During gaseous cavitation, degassing occurs because many gas bubbles move to the liquid surface by buoyancy and radiation force, and disappear at the

Fig. 1.10 Gaseous and vaporous cavitation. Reprinted with permission from Hiramatsu and Watanabe [26]. Copyright (1999), Wiley

Gaseous
cavitation

Vaporous
cavitation

liquid surface releasing gas into atmosphere. Thus, gas concentration in liquid decreases with time. It results in vaporous cavitaion that gas bubbles hardly form in liquid due to lower gas concentration. Mist formation from generated "fountain" at the liquid surface becomes much more intense. During vaporous cavitation, gas (air) dissolves into liquid from the vibrating liquid surface of the "fountain." Finally, gas concentration in the liquid becomes sufficiently high for gas-bubble formation. Then again, gaseous cavitation occurs. Repeated gaseous and vaporous cavitations have been reported for relatively high ultrasonic frequencies (500 kHz–1 MHz). Sonoluminescence as well as cavitational noise was only observed for gaseous cavitation and not for vaporous cavitation [26].

1.8 Bubble Structures

Usually, strongly pulsating active bubbles have ambient radii of a few micrometers, where the ambient bubble radius is defined as the bubble radius in the absence of ultrasound [27, 28]. Such tiny bubbles move toward a pressure antinode due to the radiation force (called primary Bjerknes force) in an ultrasonic bath when the acoustic pressure amplitude at an antinode is not too large [1, 29, 30] (see Sect. 2.16). Here, "primary" means the direct radiation force acting on a bubble from ultrasound. "Secondary" Bjerknes force is a radiation force acting between bubbles. The secondary Bjerknes force is an attractive force between tiny active bubbles. Between a tiny active bubble and a much larger bubble, it is repulsive [1, 31]. As a result of the primary and secondary Bjerknes forces, streamers of bubbles moving toward a pressure antinode are observed [32]. Near an antinode, tiny bubbles coalesce and become larger bubbles. Larger bubbles are repelled from a pressure antinode due to the nature of the primary Bjerknes force. Such large bubbles are attracted to a pressure node. Movement of cavitation bubbles due to radiation forces causes a formation of *bubble structures* [30, 33]. Some of the streaming structures are like structures of thunder. Such streamers are called "acoustic Lichtenberg figures." Lichtenberg figure is a fractal-like figure seen in electrical discharges. (Lichtenberg (1742–1799) was a German professor at the University of Goettingen.)

A type of acoustic Lichtenberg figure is shown in Fig. 1.11 at an ultrasonic frequency of 23 kHz [32]. By increasing the ultrasonic power from (a) to (f), the number of bubbles increased and the bubbles were attracted to a pressure antinode near the center of each picture. Further increase in ultrasonic power from (f) to (j) resulted in the repulsion of bubbles from a pressure antinode due to the nature of the primary Bjerknes force.

There are a variety of bubble structures in acoustic cavitation [20, 33]. Jellyfish structure is a slightly curved circular structure of bubbles in a "dendritic" form between a pressure antinode and a node. It is sometimes observed just below the liquid surface when the ultrasonic power is relatively high at relatively low ultrasonic frequencies 20–100 kHz. In other conditions, it is observed as a double layer structure: a pair of jellyfishes.

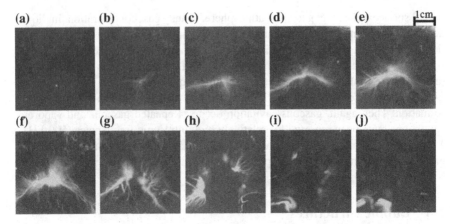

Fig. 1.11 Type of acoustic Lichtenberg figure at 23 kHz observed using a still camera with an exposure time of 2 ms and with a signal generator outputs of **a** 100, **b** 200, **c** 300, **d** 500, **e** 600, **f** 700, **g** 900, **h** 1000, **i** 1100, and **j** 1200 mVp-p, respectively. Reprinted with permission from Hatanaka et al. [32]. Copyright (2001), the Japan Society of Applied Physics

Another strange structure is a bubble cluster shown in Fig. 1.12 [34]. A cluster consisting of many bubbles was moving very fast in the liquid, similarly to a "single" bubble. The mechanism of the stability of a bubble cluster is still under

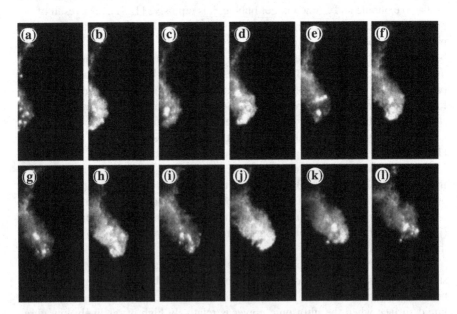

Fig. 1.12 A bubble cluster at 23 kHz with a signal generator output of 1500 mVp-p observed using a high-speed camera at 1000 fps with a 50 μs exposure. Reprinted with permission from Hatanaka et al. [34]. Copyright (2002), Elsevier

active debate. It is possible that a bubble cluster is a dynamic system; coalescence and fragmentation of bubbles repeatedly occur in a cluster [22].

1.9 Sonoluminescence

The light emission in single-bubble sonoluminescence (SBSL) in water originates from weakly ionized gases (plasma) inside a bubble at the end of a violent bubble collapse (Fig. 1.9) [5, 24]. At the end of a violent bubble collapse, temperature and pressure inside a bubble increase to 10^4 K and 10 GPa, respectively, due to a quasi-adiabatic compression of the bubble. Here, "quasi-adiabatic" means that there is considerable heat loss from a bubble due to thermal conduction to a surrounding liquid (water). The spectra of SBSL in water are mostly featureless continuum. The only exception is the OH line (310 nm in wavelength) observed from very dim SBSL [35]. The continuum emission in SBSL is associated with thermal motion of free electrons inside a heated and compressed bubble. The emission processes are electron-atom bremsstrahlung, electron-ion bremsstrahlung, and radiative recombination of electrons and ions (Fig. 1.13) [5, 24, 36–38]. When an electron is decelerated by a collision with a neutral atom (a positive ion), light is emitted, which is called electron-atom (electron-ion) bremsstrahlung. Bremsstrahlung (a German word) is electromagnetic radiation produced by a deceleration of a charged particle. Details of radiation by moving charges are described using mathematical equations in the textbook by Jackson [39]. When an electron is recombined with a positive ion, light is emitted, which is called *radiative recombination*. In common plasma, electrons interact with ions much more frequently than with neutral atoms because interaction between a neutral atom and an electron occurs only when they collide very closely. Nevertheless, in a SBSL bubble, electron-atom bremsstrahlung is probably most dominant because the density inside a bubble is nearly as high as that of condensed phase (liquid) [28].

SBSL from sulfuric acid (H_2SO_4) is over 10^3 times brighter than SBSL from water [5, 40]. It is easily observable by naked eyes even in a bright room. Due to the brightness of SBSL from sulfuric acid, analysis of the spectra is possible to quantify the intra-bubble conditions. Plasma formation inside a bubble has been confirmed by the observation of emission lines from ions such as O_2^+, Ar^+, Kr^+, and Xe^+ in the presence of Ar, Kr, and Xe, respectively, dissolved in the liquid (Fig. 1.14) [41]. O_2 is created from water vapor (H_2O). Detailed mechanism(s) of the light emission in SBSL from sulfuric acid is still under debate as well as reasons for the brightness. However, the most promising explanation is by An and Li [42] based on their numerical simulations (Figs. 1.15 and 1.16). In 85% aqueous H_2SO_4 solution, mole fraction of water vapor inside a SBSL bubble is much smaller than that inside a SBSL bubble in water according to their numerical simulations. In addition, a SBSL bubble can be driven by much intense ultrasound in aqueous H_2SO_4 solution at high concentration because the viscosity of sulfuric acid (23.8×10^{-3} Pa s at 25 °C) is more than one order of magnitude higher than that of pure water (0.89×10^{-3} Pa s

Fig. 1.13 Radiative
processes in a
sonoluminescence bubble.
a Electron-atom
bremsstrahlung,
b electron-ion
bremsstrahlung, and
c radiative recombination of
an electron and a positive ion

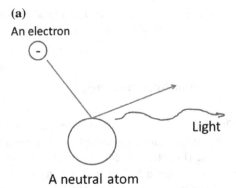

(a)

An electron

A neutral atom

Light

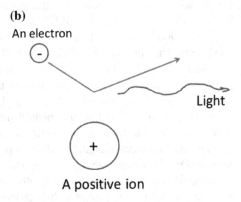

(b)

An electron

Light

A positive ion

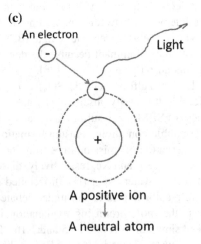

(c)

An electron   Light

A positive ion

A neutral atom

at 25 °C). Because of high viscosity of aqueous H_2SO_4 solution, a bubble is much
more shape stable compared to a bubble in pure water. As the maximum driving
pressure amplitude is determined by the shape instability of a bubble in most cases,

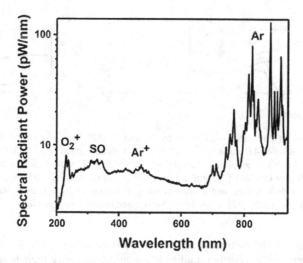

Fig. 1.14 Plasma line emission observed in a single-bubble sonoluminescence from sulfuric acid (85% H_2SO_4 in aqueous solution with 50 torr Ar). The acoustic pressure amplitude was 2.2 bar. Reprinted with permission from Flannigan and Suslick [41]. Copyright (2005), American Physical Society

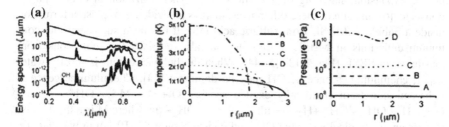

Fig. 1.15 Results of numerical simulations by An and Li [42] for an Ar bubble in 85% H_2SO_4 at 20 °C under four (4) different acoustic amplitudes of 1.5, 1.7, 2.0, and 4.0 atm marked as A, B, C, and D, respectively. The ultrasonic frequency is 37.8 kHz, and the ambient bubble radius is 13.5 μm. **a** Energy spectra of emitted light, **b** bubble temperatures at the minimum bubble radius, and **c** corresponding pressures inside a bubble. Reprinted with permission from An and Li [42]. Copyright (2009), American Physical Society

a SBSL bubble in sulfuric acid can be driven by "powerful" ultrasound. Then, the bubble temperature at the end of a violent collapse in sulfuric acid ($\sim 50,000$ K) is much higher than that in pure water ($\sim 30,000$ K). Furthermore, an equilibrium bubble radius in sulfuric acid (~ 13.5 μm) is much larger than that in pure water (~ 4 μm). Thus, the resultant sonoluminescence intensity is about 1000 times higher in sulfuric acid than that in water (Figs. 1.15a and 1.16a) [42]. The absence of line emissions in bright SBSL in pure water is due to much higher pressure inside a bubble, resulting in the broadening of spectral lines by more frequent collisions of atoms and molecules (Figs. 1.15c and 1.16c). Much higher pressure inside a SBSL

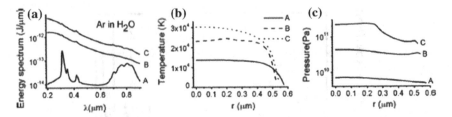

Fig. 1.16 Results of numerical simulations by An and Li [42] for an Ar bubble in water with ambient bubble radius of 4.0 μm. Curves (A) at 20 °C with ultrasound of 33.8 kHz and 1.22 atm in frequency and pressure amplitude, respectively. (B) At 20 °C with 33.8 kHz and 1.32 atm. (C) At 0 °C with 31.9 kHz and 1.32 atm. **a** Energy spectra of emitted light, **b** bubble temperatures at the minimum bubble radius, and **c** corresponding pressures inside a bubble. Reprinted with permission from An and Li [42]. Copyright (2009), American Physical Society

bubble in pure water compared to that in sulfuric acid is due to much lower viscosity of pure water. Further studies are required to verify their findings [42].

SBSL occurs from partially degassed (undersaturated) water because many bubbles are easily created in liquid saturated with gas. Light emissions from many bubbles are called multibubble sonoluminescence (MBSL). In this case, there are bubble–bubble interactions which may result in jetting into bubbles, suppression of bubble expansion, shielding of acoustic waves, synchronization in bubble pulsations, etc. [6]. In spite of these differences, there is an evidence of plasma formation inside a bubble in MBSL from sulfuric acid [43]. It has been suggested that continuum emissions in MBSL from water are also originated in emissions from plasma as in SBSL (Fig. 1.13) [5, 44]. When the bubble content is mostly water vapor, chemiluminescence of OH may be dominant [44]. Chemiluminescence is light emission from chemically excited species. $O + H + M \rightarrow OH^* + M$ and $OH + H + OH \rightarrow OH^* + H_2O$ result in $OH^* \rightarrow OH + h\nu$, where M is a third body, OH^* is an electronically excited OH, and $h\nu$ is a photon of 310 nm in wavelength. However, detailed mechanism(s) of OH line emission in sonoluminescence is still unclear (see Sect. 3.6).

Difference and similarity between SBSL and MBSL are still under intense debate. Most significant difference between SBSL and MBSL is the absence and presence of alkali-metal emission lines, respectively, in aqueous alkali-metal solutions. For example, in 0.1 M sodium chloride (NaCl) solution, Na-line emission at about 590 nm in wavelength was absent and present in SBSL and MBSL, respectively (Fig. 1.17) [45]. However, in SBSL from 74% H_2SO_4 with 1% Na_2SO_4, Na-line was observed at very high acoustic pressure amplitude such as 5 bar (Fig. 1.18) [46]. In this case, a SBSL bubble was moving around the pressure antinode. It is suggested that shape instability of a moving bubble results in Na-line emission. In aqueous solution, Na is mostly present as an ion (Na^+). As an ion is nonvolatile, Na atoms do not enter the interior of a bubble. In order to excite Na atoms to emit Na-line, there are only two possibilities. One is that Na atoms are thermally (or chemically) excited in the heated interior of a bubble. For this

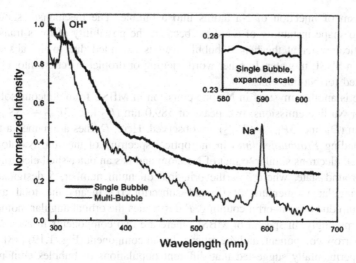

Fig. 1.17 Comparison of the background subtracted spectra of MBSL and SBSL in a 0.1 M sodium chloride (NaCl) solution. Each spectrum was normalized to its highest intensity. Reprinted with permission from Matula et al. [45]. Copyright (1995), American Physical Society

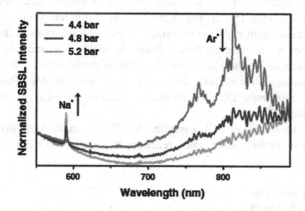

Fig. 1.18 SBSL spectra from a moving bubble at around the pressure antinode in 74% H_2SO_4 with 1% Na_2SO_4 aqueous solution re-gassed with 50 torr Ar. The spectra have been normalized to SL intensity at 500 nm. The arrows next to the spectral features indicate how the intensities of the corresponding features change with increasing acoustic pressure amplitude. Reprinted with permission from Flannigan and Suslick [46]. Copyright (2007), American Physical Society

possibility, Na atoms should be injected into a bubble by jetting, mist (tiny droplets) formation, or mass transfer of supercritical water at the heated bubble wall. The other is that Na atoms are thermally (or chemically) excited in liquid outside a bubble. As Na-line intensity and its detailed spectral shape both depend on the gas content in a bubble in experiments, it has been suggested that Na-line is actually emitted in the interior of a heated bubble [47]. Then, there should be some mechanism of injection of Na atoms into a bubble. The mechanism should be related to shape instability of a bubble because the possibility of mass transfer of supercritical water at the heated bubble wall is excluded due to the absence of Na-line in SBSL in water. In other words, jetting or droplet injection into a bubble is required for Na-line emission.

There is another mystery in Na-line emission in MBSL from aqueous solutions [48]. For Na-line emissions, two peaks at 589.0 nm (D_2 line, $3P_{3/2} \rightarrow 3S_{1/2}$) and 589.6 nm ($D_1$ line, $3P_{1/2} \rightarrow 3S_{1/2}$) are observed. Here, D lines are named after the corresponding *Fraunhofer lines* in the optical spectrum of the sun. Na atom has inner-shell electrons similar to that of Ne atom and has an outer-shell electron $3S_{1/2}$ at the ground state, where 3 is the principal quantum number, S designates the orbital angular momentum of 0, and subscript 1/2 means the total angular momentum due to spin–orbit coupling. P designates the orbital angular momentum of 1. Surprisingly, in spectra of MBSL, there are two components in each D line. One is narrow component, and the other is broad component (Fig. 1.19) [48]. It has been experimentally suggested that different populations of bubbles emit narrow and broad components of D lines. Hotter bubbles emit narrow component, and colder bubbles emit broad component. Further studies are required on this topic.

Relatively bright emission of chemiluminescence of luminol (430 nm in wavelength) in aqueous alkaline solution irradiated by intense ultrasound is called sonochemiluminescence (SCL) (Fig. 1.20) [49]. It is different from sonoluminescence (SL) because light emission originates in chemical reactions of luminol with oxidants created by acoustic cavitation. Oxidants such as OH radicals (OH·) and H_2O_2 are created inside the heated bubbles. The oxidants diffuse out of a bubble into the surrounding liquid and chemically react with solutes. This is called *sonochemical reactions*. Detailed chemical reactions in SCL of luminol are described in Ref. [50].

The use of lucigenin for SCL experiments (528 nm in wavelength) instead of luminol is due to the fact that chemical reactions of lucigenin with reductive species

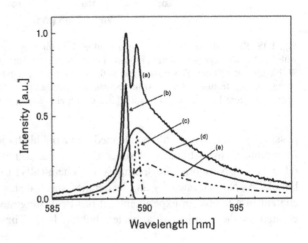

Fig. 1.19 Separation of the spectrum of Na emission in MBSL from Ar-saturated 4 M NaCl aqueous solution irradiated with ultrasound (145 kHz and 12 W). (a) Observed spectrum. (b) Narrow component of D_2 line. (c) Narrow component of D_1 line. (d) Broad component of D_2 line. (e) Broad component of D_1 line. Reprinted with permission from Nakajima et al. [48]. Copyright (2015), The Japan Society of Applied Physics

Fig. 1.20 A photograph of
SCL from 1 mM sodium
carbonate-0.01 mM luminol
aqueous solution irradiated
with ultrasound (140 kHz)
from the bottom of a liquid
container. The signal
generator output was 300
mVp-p. The exposure time for
the photograph was 20 min.
using a film of ISO 1600.
Reprinted with permission
from Hatanaka et al. [49].
Copyright (2000), the Japan
Society of Applied Physics

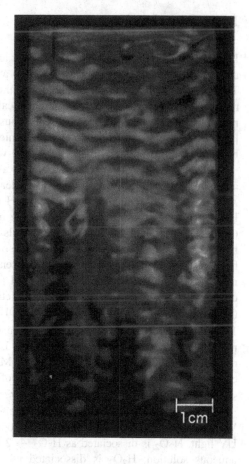

easily occur in an ultrasonic reaction field [51, 52]. Reductive species are often
generated by the reaction of OH radicals and organic material such as alcohol added
in aqueous solution as follows.

$$OH + CH_3CH(OH) CH_3 \rightarrow CH_3C(OH)CH_3 + H_2O \qquad (1.30)$$

where abstraction of H atom from C–H bonds leads to a 2-propanol radical in the
right side. The 2-propanol radical is a reductive radical with a reduction potential of
−1.39 V. A reductive radical reacts with O_2 and lucigenin (Luc^{2+}) as follows.

$$CH_3C(OH) CH_3 + O_2 \rightarrow O_2^- + CH_3COCH_3 + H^+ \qquad (1.31)$$

$$CH_3C(OH)CH_3 + Luc^{2+} \rightarrow Luc^+ + CH_3COCH_3 + H^+ \qquad (1.32)$$

A lucigenin cation radical (Luc +) reacts with superoxide radical (O_2^-) and emits
chemiluminescence light.

1.10 Sonochemistry

In a heated interior of a bubble, water vapor and oxygen, if present, are dissociated and oxidants such as OH radicals, H_2O_2, O atoms, and O_3 are formed (Fig. 1.21). The oxidants diffuse out of a bubble into the surrounding liquid and chemically react with solutes if present. Such chemical reactions are called sonochemical reactions, and chemistry associated with acoustic cavitation is called *sonochemistry* [53]. Most dominant oxidant in sonochemical reactions is usually OH radicals because O atoms probably react with liquid water at the bubble wall as O· + H_2O → H_2O_2 and O_3 is not produced as much as OH· [23]. However, the role of O atoms in sonochemical reactions is still unclear as described in Sects. 2.10 and 3.9. As the oxidation–reduction potential of OH radicals is much higher than that of H_2O_2, OH often plays more important role in sonochemical reactions than H_2O_2 (Table 1.1) [54]. The lifetime of OH radicals in liquid water is mainly determined by the reaction between them in the absence of solutes: OH· + OH· → H_2O_2. It has been experimentally suggested that the concentration of OH radicals in liquid water near the bubble wall is about 5×10^{-3} M (= mol/l) [55]. When the initial concentration of OH· is 5×10^{-3} M, the lifetime of OH radicals is about 20 ns (Table 1.2) [56]. The diffusion length for OH radicals during the lifetime is about 13 nm, which is calculated by $2\sqrt{Dt}$ where D is the diffusion coefficient for OH radicals (= 2.2×10^{-9} m^2/s at 25 °C) and t is time. The rate constants of reactions of OH· with solutes are typically 10^7–10^{10} M^{-1} s^{-1} (Table 1.2) [56]. When solute concentration is more than 0.05 M, lifetime of OH radicals is mainly determined by solute concentration if the rate constant is ≈ 10^9 M^{-1}s^{-1}.

The lifetime of H_2O_2 is long in the absence of solutes, UV light, and catalyst: With catalyst such as MnO_2, H_2O_2 is dissociated as $2H_2O_2$ → $2H_2O$ + O_2; with UV light, H_2O_2 is dissociated as H_2O_2 → 2 OH· [57]; and with Fe^{2+} (or Cu^+) in aqueous solution, H_2O_2 is dissociated as Fe^{2+} + H_2O_2 → Fe^{3+} + OH· + OH$^-$ (Fenton reaction).

Fig. 1.21 Production of oxidants inside a bubble in water irradiated with power ultrasound (sonochemical reactions)

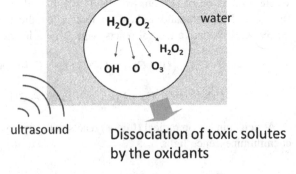

Production of oxidants

A bubble

H_2O, O_2 water

H_2O_2

OH   O O_3

ultrasound   Dissociation of toxic solutes by the oxidants

Table 1.1 Oxidation–reduction potential of typical oxidants [54]

Reaction	Potential (V)
$O\cdot (g) + 2H^+ + 2e^- \rightarrow H_2O$	2.421
$O_3 + 2H^+ + 2e^- \rightarrow O_2 + H_2O$	2.076
$OH\cdot + e^- \rightarrow OH^-$	2.02
$H_2O_2 + 2H^+ + 2e^- \rightarrow 2H_2O$	1.776
$HO_2 + H^+ + e^- \rightarrow H_2O_2$	1.495

Table 1.2 Rate constants of reactions of OH radicals in aqueous solution at 25 °C (M = mol/l) [56]

Reaction	Rate constant ($M^{-1}\ s^{-1}$)
$OH\cdot + OH\cdot \rightarrow H_2O_2$	100×10^8
$OH\cdot + H_2O_2 \rightarrow H_2O + HO_2$	0.3×10^8
$OH\cdot + HCO_2^- \rightarrow CO_2^- + H_2O$	65×10^8
$OH\cdot + \text{methanol} \rightarrow \text{products}$	1.1×10^8
$OH\cdot + \text{2-propanol} \rightarrow \text{products}$	29×10^8
$OH\cdot + Fe(CN)_6^{4-} \rightarrow Fe(CN)_6^{3-} + OH^-$	400×10^8

The first direct evidence of OH radical production in acoustic cavitation was obtained by ESR spectra of spin-trapped radicals from argon-saturated aqueous solutions containing DMPO (the nitrone spin trap) irradiated by ultrasound (50 kHz, ~0.06 W/cm²) by Makino, Mossoba, and Riesz in 1982 (Fig. 1.22) [58]. ESR signals from H radicals (H·) (H-DMPO adduct) were also observed in addition

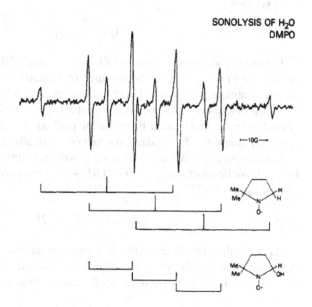

Fig. 1.22 ESR spectrum of an argon-saturated aqueous DMPO solution (25 mM) irradiated with ultrasound (50 kHz) for 3 min. The spectrum consists of OH and H adducts as indicated by the stick diagrams, implying that OH and H radicals are created by the sonolysis of water. Reprinted with permission from Makino et al. [58]. Copyright (1982), American Chemical Society

SONOLYSIS OF H₂O
DMPO

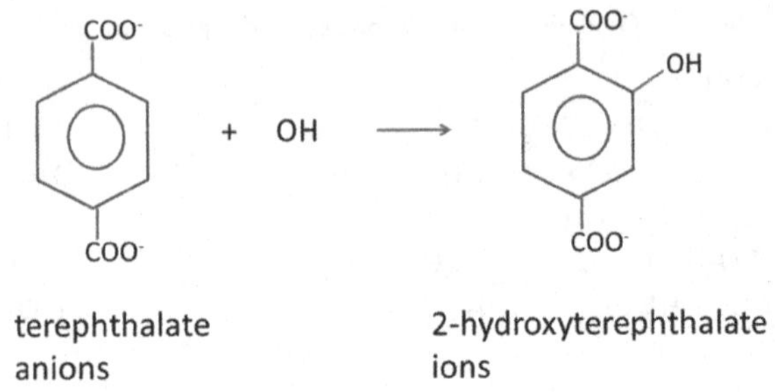

Fig. 1.23 Terephthalate
dosimetry

terephthalate
anions

2-hydroxyterephthalate
ions

to that from OH radicals (OH-DMPO adduct). The same ESR signals as those from
OH-DMPO adduct are, however, observed in the presence of UV light, H_2O_2, or
metal ions such as Fe^{3+} [59, 60]. Thus, Makino et al. [58, 61] confirmed the
production of OH and H radicals by the observation of the decrease in the ESR
signals by adding OH and H scavengers such as methanol, ethanol, acetone.

More direct evidence of OH radical production in acoustic cavitation has been
obtained by terephthalate dosimetry (Fig. 1.23) [62–64]. In an alkaline aqueous
solution, terephthalic acid produces terephthalate anions that react with OH radicals
to generate highly fluorescent 2-hydroxyterephthalate ions.

A standard method to quantify the amount of oxidants produced in acoustic
cavitation is the potassium iodide (KI) dosimetry [65]. In an aqueous KI solution, I^-
ions are oxidized to give I_2.

$$2OH \cdot + 2I^- \rightarrow 2OH^- + I_2 \tag{1.33}$$

When excess I^- ions are present in solutions, I_2 reacts with the excess I^- ions to
form I_3^- ions.

$$I_2 + I^- \rightarrow I_3^- \tag{1.34}$$

For such experiments, a standard KI concentration of 0.1 M is usually used. The
absorbance of I_3^- at 355 nm is measured (e = 26303 M^{-1} cm^{-1}). An example of
the measurement is shown in Fig. 1.24 as a function of ultrasonic frequency [65].
The ultrasonic power was measured by calorimetry. Typical average concentration
of oxidants produced in acoustic cavitation per hour was about 10 μM. The optimal
ultrasonic frequency for oxidants production is usually 200–500 kHz.

In radiation chemistry using ionizing radiation, hydrogen atoms are also formed
by the dissociation of water: $H_2O \rightarrow OH \cdot + H \cdot$. Hydrogen atoms reduce the iodine
formed as follows [55].

$$2H \cdot + I_2 \rightarrow 2I^- + 2H^+ \tag{1.35}$$

As a result, very little iodine is formed in radiation chemistry. In acoustic
cavitation (sonochemistry), on the other hand, the reaction (1.35) is minor because
the amount of H production is much smaller than that of OH radicals [55].

Fig. 1.24 Frequency dependence of KI oxidation yield per unit ultrasonic power. The ultrasonic power was measured by calorimetry. KI concentration was 0.1 M. The liquid temperature was about 25 °C. Reprinted with permission from Koda et al. [65]. Copyright (2003), Elsevier

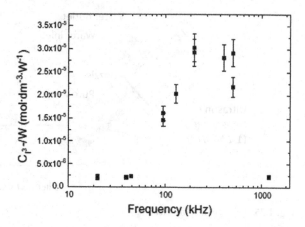

According to the numerical simulations of chemical reactions inside a heated bubble, this is due to the fact that H radicals are consumed: $H_2O + H\cdot \rightarrow OH\cdot + H_2$, $HO_2 + H\cdot \rightarrow 2 OH\cdot$, and $2H\cdot \rightarrow H_2$ [66]. The production of OH radicals inside a heated bubble is due to the following reactions: $H_2O + M \rightarrow OH\cdot + H\cdot + M$ (M: a third body), $H_2O + O\cdot \rightarrow 2OH\cdot$, $H_2O + H\cdot \rightarrow OH\cdot + H_2$, and $HO_2 + H\cdot \rightarrow 2 OH\cdot$ [66].

There are many other chemical effects induced by power ultrasound. One is the reduction in liquid viscosity by acoustic cavitation [67]. For example, polymer chains are often cut by violently pulsating bubbles [68]. Another is the enhancement in crystal nucleation which is called sonocrystallization [69, 70]. However, the detailed mechanism for sonocrystallization is still under debate as described in Sect. 3.3. Another example is the dissolution of gels into liquid by acoustic cavitation, resulting in different mechanisms of nanoparticle formation compared to simple stirring [71]. It has been reported that sonochemically synthesized nanocrystals are often mesocrystals which are aggregates of nanocrystals with their crystal axes aligned [72, 73].

1.11 Ultrasonic Cleaning

There are typically two types of equipment for ultrasonic cleaning. One is an ultrasonic bath in which samples are immersed into the liquid. Typical ultrasonic frequency for a bath is around 40 kHz. For milder cleaning, sometimes ultrasonic waves with higher frequencies than 1 MHz are employed (megasonic cleaning). For lower ultrasonic frequencies, cleaning due to acoustic cavitation is more intense. However, damage (or erosion) due to acoustic cavitation is also increased. Another problem in ultrasonic bath type is the frequent reattachment of removed particles (contaminations) to samples to be cleaned. Frequency of reattachment is

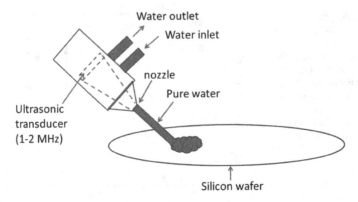

Fig. 1.25 Ultrasonic spray cleaning

dramatically reduced by using ultrasonic spray (Fig. 1.25). In ultrasonic spray cleaning, contaminants are moved away with the liquid flow associated with spraying.

There are hundreds of processes to produce integrated circuits [IC or LSI (large-scaled IC)] on a silicon wafer (a thin plate of a single crystal of Si) such as printing of LSI circuit patterns on a silicon wafer using a method similar to photography, dicing to cut a wafer as chips. About 20% of the processes are related to cleaning, and more than 50% of defective products are due to particle contamination on printed circuits which cause defects in circuit patterns (a short circuit, etc.) and/or deterioration of device properties.

Before 2000, cleaning of silicon wafers was conducted by the conventional RCA cleaning developed by RCA cooperation in USA in 1970. However, the conventional RCA cleaning had a problem in that concentration of solutes in aqueous solutions changed with time due to evaporation because aqueous solutions of NH_4OH and H_2O_2 were used at temperatures higher than 80 °C. In addition, a large amount of harmful chemicals were employed in cleaning processes, which increased environmental and health risks. In order to decrease those risks, new cleaning (mechanical) methods such as brush, ultrasonics, jets were implemented in many industries. For example, a cleaning method developed by Ohmi [74] in 1996 successfully decreased risks by using ultrasonic cleaning in pure water containing HF, H_2O_2 (≤ 1 %) and containing O_3 (1 ppm) at room temperature (Fig. 1.26).

The main mechanism of ultrasonic cleaning is mostly a physical effect at relatively low ultrasonic frequencies [6]. Details will be discussed in Sect. 2.15. At relatively high ultrasonic frequencies, acoustic streaming may also contribute to cleaning [75]. However, there are two problems when using ultrasonic cleaning. One is the nonuniformity of cleaning, and the other is the physical damage (mainly erosion). To prevent this physical damage, a variety of methods have been proposed. One possibility is to use very high ultrasonic frequencies, i.e., in the GHz region (= 10^9 Hz = 1000 MHz) called gigasonic [76]. This is due to the fact that physical effects induced by acoustic cavitation become weaker as the ultrasonic

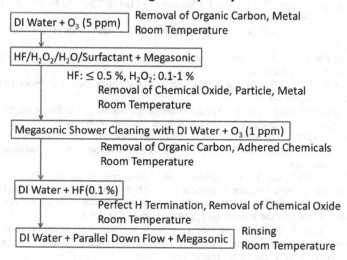

Wet Cleaning developed by Ohmi

DI Water + O₃ (5 ppm) — Removal of Organic Carbon, Metal
Room Temperature

HF/H₂O₂/H₂O/Surfactant + Megasonic
HF: ≤ 0.5 %, H₂O₂: 0.1-1 %
Removal of Chemical Oxide, Particle, Metal
Room Temperature

Megasonic Shower Cleaning with DI Water + O₃ (1 ppm)
Removal of Organic Carbon, Adhered Chemicals
Room Temperature

DI Water + HF(0.1 %)
Perfect H Termination, Removal of Chemical Oxide
Room Temperature

DI Water + Parallel Down Flow + Megasonic — Rinsing
Room Temperature

Fig. 1.26 Wet cleaning developed by Ohmi in 1996 [74]

frequency increases. Another possibility is to use microbubbles produced by a microbubble generator in conjunction with power ultrasound [77, 78]. In this set-up, microbubbles are generated by hydrodynamic cavitation and ultrasound is strongly attenuated. Furthermore, bubble collapse becomes milder due to the bubble–bubble interaction [79, 80]. For bubble–bubble interaction, we discuss in Sect. 2.18.

References

1. Leighton TG (1994) The acoustic bubble. Academic Press, London
2. Fahy F (2001) Foundations of engineering acoustics. Academic Press, San Diego
3. Pierce AD (1989) Acoustics, an introduction to its physical principles and applications. Acoustical Society of America, New York
4. Maris H, Balibar S (2000) Negative pressures and cavitation in liquid helium. Phys Today 53 (2):29–34. doi:10.10631/1.882962
5. Yasui K, Tuziuti T, Sivakumar M, Iida Y (2004) Sonoluminescence. Appl Spectrosc Rev 39:399–436. doi:10.1081/ASR-200030202
6. Yasui K (2015) Dynamics of acoustic bubbles. In: Grieser F, Choi PK, Enomoto N, Harada H, Okitsu K, Yasui K (eds) Sonochemistry and the acoustic bubble. Elsevier, Amsterdam
7. Galloway WJ (1954) An experimental study of acoustically induced cavitation in liquids. J Acoust Soc Am 26:849–857. doi:10.1121/1.1907428
8. Yasui K (2002) Influence of ultrasonic frequency on multibubble sonoluminescence. J Acoust Soc Am 112:1405–1413. doi:10.1121/1.1502898

9. Apfel RE, Holland CK (1991) Gauging the likelihood of cavitation from short-pulse, low-duty cycle diagnostic ultrasound. Ultrasound in Med & Biol 17:179–185. doi:10.1016/0301-5629(91)90125-G
10. Kinsler LE, Frey AR, Coppens AB, Sanders JV (1982) Fundamentals of acoustics, 3rd edn. Wiley, New York
11. Kremkau FW (2006) Diagnostic ultrasound: principles and instruments, 7th edn. Saunders Elsevier, St. Louis, Missouri
12. Ter Haar G (2011) Ultrasonic imaging: safety considerations. Interface Focus 1:686–697. doi:10.1098/rsfs.2011.0029
13. Wu J, Nyborg W (eds) (2006) Emerging therapeutic ultrasound. World Scientific, New Jersey
14. Kittel C (2005) Introduction to solid state physics, 8th edn. Wiley, New York
15. Asakura Y (2015) Experimental methods in sonochemistry. In: Grieser F, Choi PK, Enomoto N, Harada H, Okitsu K, Yasui K (eds) Sonochemistry and the acoustic bubble. Elsevier, Amsterdam
16. Yasui K, Tuziuti T, Iida Y (2005) Dependence of the characteristics of bubbles on types of sonochemical reactors. Ultrason Sonochem 12:43–51. doi:10.1016/j.ultaonch.2004.06.003
17. Hacias KJ, Cormier GJ, Nourie SM, Kubel EJ Jr (1997) Guide to acid, alkaline, emulsion, and ultrasonic cleaning. ASM International, Materials Park, OH, USA
18. Tuziuti T, Yasui K, Sivakumar M, Iida Y, Miyoshi N (2005) Correlation between acoustic cavitation noise and yield enhancement of sonochemical reaction by particle addition. J Phys Chem 109:4869–4872. doi:10.1021/jp0503516
19. Yasui K (2011) Fundamentals of acoustic cavitation and sonochemistry. In: Pankaj Ashokkumar M (ed) Theoretical and experimental sonochemistry involving inorganic systems. Springer, Dordrecht
20. Yasui K, Izu N (2017) Effect of evaporation and condensation on a thermoacoustic engine: a Lagrangian simulation approach. J Acoust Soc Am 141:4398–4407. doi:10.1121/1.4985385
21. Beyer RT (1997) Nonlinear acoustics. Acoustical Society of America, New York
22. Yasui K, Iida Y, Tuziuti T, Kozuka T, Towata A (2008) Strongly interacting bubbles under an ultrasonic horn. Phys Rev E 77:016609. doi:10.1103/PhysRevE.77.016609
23. Yasui K (2016) Unsolved problems in acoustic cavitation. In: Ashokkumar M, Cavalieri F, Chemat F, Okitsu K, Sambandam A, Yasui K, Zisu B (eds) Handbook of ultrasonics and sonochemistry. Springer, Singapore
24. Brenner MP, Hilgenfeldt S, Lohse D (2002) Single-bubble sonoluminescence. Rev Mod Phys 74:425–484. doi:10.1103/RevModPhys.74.425
25. Degrois M (1966) Cavitation oscillation. Ultrasonics 4:38–39. doi:10.1016/0041-624X(66)90012-6
26. Hiramatsu S, Watanabe Y (1999) On the mechanism of relaxation oscillation in sonoluminescence. Electro Commun Jpn Part 3 82(2):58–65. doi:10.1002/(SICI)1520-6440(199902)82:2<58::AID-ECJC7>3.0.CO;2-#
27. Weninger KR, Camara CG, Putterman SJ (2001) Observation of bubble dynamics within luminescent cavitation clouds: sonoluminescence at the nano-scale. Phys Rev E 63:016310. doi:10.1103/PhysRevE.63.016310
28. Yasui K, Tuziuti T, Lee J, Kozuka T, Towata A, Iida Y (2008) The range of ambient radius for an active bubble in sonoluminescence and sonochemical reactions. J Chem Phys 128:184705. doi:10.1063/1.2919119
29. Matula TJ, Cordry SM, Roy RA, Crum LA (1997) Bjerknes force and bubble levitation under single-bubble sonoluminescence conditions. J Acoust Soc Am 102:1522–1527. doi:10.1121/1.420065
30. Mettin R (2007) From a single bubble to bubble structures in acoustic cavitation. In: Kurz T, Parlitz U, Kaatze U (eds) Oscillations, waves and interactions. Universitatsverlag Goettingen, Goettingen
31. Mettin R, Cairos C (2016) Bubble dynamics and observations. In: Ashokkumar M, Cavalieri F, Chemat F, Okitsu K, Sambandam A, Yasui K, Zisu B (eds) Handbook of ultrasonics and sonochemistry. Springer, Singapore

32. Hatanaka S, Yasui K, Tuziuti T, Kozuka T, Mitome H (2001) Quenching mechanism of multibubble sonoluminescence at excessive sound pressure. Jpn J Appl Phys 40:3856–3860. doi:10.1143/JJAP.40.3856

33. Mettin R (2005) Bubble structures in acoustic cavitation. In: Doinikov AA (ed) Bubble and particle dynamics in acoustic fields: modern trends and applications. Research Signpost, Kerala, India

34. Hatanaka S, Yasui K, Kozuka T, Tuziuti T, Mitome H (2002) Influence of bubble clustering on multibubble sonoluminescence. Ultrasonics 40:655–660. doi:10.1016/S0041-624X(02) 00193-2

35. Young JB, Nelson JA, Kang W (2001) Line emission in single-bubble sonoluminescence. Phys Rev Lett 86:2673–2676. doi:10.1103/PhysRevLett.86.2673

36. Hilgenfeldt S, Grossmann S, Lohse D (1999) A simple explanation of light emission in sonoluminescence. Nature (London) 398:402–405

37. Hilgenfeldt S, Grossmann S, Lohse D (1999) Sonoluminescence light emission. Phys Fluids 11:1318–1330. doi:10.1063/1.869997

38. Yasui K (1999) Mechanism of single-bubble sonoluminescence. Phys Rev E 60:1754–1758. doi:10.1103/PhysRevE.60.1754

39. Jackson JD (1975) Classical electrodynamics, 2nd edn. Wiley, New York

40. Suslick KS, Flannigan DJ (2008) Inside a collapsing bubble: sonoluminescence and the conditions during cavitation. Annu Rev Phys Chem 59:659–683. doi:10.1146/annurev. physchem.59.032607.093739

41. Flannigan DJ, Suslick KS (2005) Plasma line emission during single-bubble cavitation. Phys Rev Lett 95:044301. doi:10.1103/PhysRevLett.95.044301

42. An Y, Li C (2009) Diagnosing temperature change inside sonoluminescing bubbles by calculating line spectra. Phys Rev E 80:046320. doi:10.1103/PhysRevE.80.046320

43. Eddingsaas NC, Suslick KS (2007) Evidence for a plasma core during multibubble sonoluminescence in sulfuric acid. J Am Chem Soc 129:3838–3839. doi:10.1021/ja070192z

44. Yasui K (2001) Temperature in multibubble sonoluminescence. J Chem Phys 115: 2893–2896. doi:10.1063/1.1395056

45. Matula TJ, Roy RA, Mourad PD, McNamara WB III, Suslick KS (1995) Comparison of multibubble and single-bubble sonoluminescence spectra. Phys Rev Lett 75:2602–2605. doi:10.1103/PhysRevLett.75.2602

46. Flannigan DJ, Suslick KS (2007) Emission from electronically excited metal atoms during single-bubble sonoluminescence. Phys Rev Lett 99:134301. doi:10.1103/PhysRevLett.99. 134301

47. Choi PK (2011) Sonoluminescence of inorganic ions in aqueous solutions. In: Pankaj, Ashokkumar M (eds) Theoretical and experimental sonochemistry involving inorganic systems. Springer, Dordrecht

48. Nakajima R, Hayashi Y, Choi PK (2015) Mechanism of two types of Na emission observed in sonoluminescence. Jpn J Appl Phys 54: 07HE02. doi:10.7567/JJAP.54.07HE02

49. Hatanaka S, Yasui K, Tuziuti T, Mitome H (2000) Difference in threshold between sono- and sonochemical luminescence. Jpn J Appl Phys 39:2962–2966. doi:10.1143/JJAP.39.2962

50. McMurray HN, Wilson BP (1999) Mechanism and spatial study of ultrasonically induced luminol chemiluminescence. J Phys Chem A 103:3955–3962. doi:10.1021/jp984503r

51. Matsuoka M, Jin J (2015) Sonochemiluminescence from lucigenin in an aqueous solution using alcohols as coreactant. Chem Lett 44:1759–1761. doi:10.1246/cl.150838

52. Matsuoka M, Takahashi F, Asakura Y, Jin J (2016) Sonochemiluminescence of lucigenin: evidence of superoxide radical anion formation by ultrasonic irradiation. Jpn J Appl Phys 55: 07KB01. doi:10.7567/JJAP.55.07KB01

53. Grieser F, Choi PK, Enomoto N, Harada H, Okitsu K, Yasui K (eds) (2015) Sonochemistry and the acoustic bubble. Elsevier, Amsterdam

54. Lide DR (ed) (1994) CRC handbook of chemistry and physics, 75th edn. CRC Press, Boca Raton

55. Henglein A (1993) Contributions to various aspects of cavitation chemistry. In: Mason TJ (ed) Advances in sonochemsitry, vol 3. JAI Press, London
56. Elliot AJ, McCracken DR, Buxton GV, Wood ND (1990) Estimation of rate constants for near-diffusion-controlled reactions in water at high temperatures. J Chem Soc, Faraday Trans 86:1539–1547. doi:10.1039/ft9908601539
57. Mugnai A, Petroncelli P, Fiocco G (1979) Sensitivity of the photodissociation of NO_2, NO_3, HNO_3 and H_2O_2 to the solar radiation diffused by the ground and by atmospheric particles. J Atmosph Terrest Phys 41:351–359. doi:10.1016/0021-9169(79)90031-X
58. Makino K, Mossoba MM, Riesz P (1982) Chemical effects of ultrasound on aqueous solutions. evidence for OH and H by spin trapping. J Am Chem Soc 104:3537–3539. doi:10.1021/ja00376a064
59. Finkelstein E, Rosen GM, Rauckman EJ (1980) Spin trapping of superoxide and hydroxyl radical: practical aspects. Archives Biochem Biophys 200:1–16. doi:10.1016/0003-9861(80)90323-9
60. Riesz P, Berdahl D, Christman CL (1985) Free radical generation by ultrasound in aqueous and nonaqueous solutions. Environ Health Perspect 64:233–252. doi:10.2307/3430013
61. Makino K, Mossoba MM, Riesz P (1983) Chemical effects of ultrasound on aqueous solutions. Formation of hydroxyl radicals and hydrogen atoms. J Phys Chem 87:1369–1377. doi:10.1021/j100231a020
62. Fang X, Mark G, von Sonntag C (1996) OH radicals formation by ultrasound in aqueous solutions part I: the chemistry underlying the terephthalate dosimeter. Ultrason Sonochem 3:57–63. doi:10.1016/1350-4177(95)00032-1
63. Mark G, Tauber A, Laupert R, Schuchmann HP, Schulz D, Mues A, von Sonntag C (1998) OH-radical formation by ultrasound in aqueous solution—part II: terephthalate and Fricke dosimetry and the influence of various conditions on the sonolytic yield. Ultrason Sonochem 5:41–52. doi:10.1016/S1350-4177(98)00012-1
64. Iida Y, Yasui K, Tuziuti T, Sivakumar M (2005) Sonochemistry and its dosimetry. Microchem J 80:159–164. doi:10.1016/j.microc.2004.07.016
65. Koda S, Kimura T, Kondo T, Mitome H (2003) A standard method to calibrate sonochemical efficiency of an individual reaction system. Ultrason Sonochem 10:149–156. doi:10.1016/S1350-4177(03)00084-1
66. Yasui K, Tuziuti T, Sivakumar M, Iida Y (2005) Theoretical study of single-bubble sonoluminescence. J Chem Phys 122:224706. doi:10.1063/1.1925607
67. Iida Y, Tuziuti T, Yasui K, Towata A, Kozuka T (2008) Control of viscosity in starch and polysaccharide solutions with ultrasound after gelatinization. Innov Food Sci Emerg Technol 9:140–146. doi:10.1016/j.ifset.2007.03.029
68. Price GJ (1990) The use of ultrasound for the controlled degradation of polymer solutions. In: Mason TJ (ed) Advances in sonochemistry, vol 1. JAO Press, Greenwich, Connecticut
69. Zhang Z, Sun DW, Zhu Z, Cheng L (2015) Enhancement of crystallization processes by power ultrasound: current state-of-the-art and research advances. Comprehensive Rev Food Sci Food Safety 14:303–316. doi:10.1111/1541-4337.12132
70. Castillo-Peinado LS, Dolores M, Castro L (2016) The role of ultrasound in pharmaceutical production: sonocrystallization. J Pharm Pharmacol 68:1249–1267. doi:10.1111/jphp.12614
71. Yasui K, Kato K (2017) Numerical simulations of sonochemical production and oriented aggregation of $BaTiO_3$ nanocrystals. Ultrason Sonochem 35:673–680. doi:10.1016/j.ultsonch.2016.05.009
72. Dang F, Kato K, Imai H, Wada S, Haneda H, Kuwabara M (2010) A new effect of ultrasonication on the formation of $BaTiO_3$ nanoparticles. Ultrason Sonochem 17:310–314. doi:10.1016/j.ultsonch.2009.08.006
73. Yasui K, Kato K (2014) Numerical simulations of nucleation and aggregation of $BaTiO_3$ nanocrystals under ultrasound. In: Manickam S, Ashokkumar M (eds) Cavitaion a novel energy-efficient technique for the generation of nanomaterials. Pan Stanford, Singapore
74. Ohmi T (1996) Total room temperature wet cleaning for Si substrate surface. J Electrochem Soc 143:2957–2964. doi:10.1149/1.1837133

75. Bakhtari K, Guldiken RO, Busnaina AA, Park JG (2006) Experimental and analytical study of submicrometer particle removal from deep trenches. J Electrochem Soc 153:C603–C607. doi:10.1149/1.2214531
76. Potter G, Tokranova N, Rastegar A, Castracane J (2016) Design, fabrication, and testing of surface acoustic wave devices for semiconductor cleaning applications. Microelectro Eng 162:100–104. doi:10.1016/j.mee.2016.04.006
77. Tuziuti T (2016) Influence of sonication conditions on the efficiency of ultrasonic cleaning with flowing micrometer-sized air bubbles. Ultrason Sonochem 29:604–611. doi:10.1016/j.ultsonch.2015.09.011
78. Iizuka A, Iwata W, Shimata E, Nakamura T (2016) Physical washing method for press oil removal from side surfaces using microbubbles under ultrasonic irradiation. Ind Eng Chem Res 55:10782–10787. doi:10.1021/acs.iecr.6b01887
79. Yasui K, Lee J, Tuziuti T, Towata A, Kozuka T, Iida Y (2009) Influence of the bubble-bubble interaction on destruction of encapsulated microbubbles under ultrasound. J Acoust Soc Am 126:973–982. doi:10.1121/1.3179677
80. Yasui K, Towata A, Tuziuti T, Kozuka T, Kato K (2011) Effect of static pressure on acoustic energy radiated by cavitation bubbles in viscous liquids under ultrasound. J Acoust Soc Am 130:3233–3242. doi:10.1121/1.3626130

Chapter 2
Bubble Dynamics

Abstract Bubble pulsation is mathematically described by the Rayleigh–Plesset equation and by Keller equation. Derivation of the equations is fully described herein. Using the Rayleigh–Plesset equation, the violent collapse of a bubble is discussed. A method of numerical simulations of bubble pulsation is also described. In relation to numerical simulations, non-equilibrium evaporation and condensation of water vapor at the bubble wall, the variation in liquid temperature at the bubble wall, the gas diffusion across the bubble wall, and the chemical reactions inside a bubble are discussed. Comparison between numerical results and experimental data for a single-bubble system is shown. The main oxidants created inside a bubble are described based upon numerical simulations data. Linear and nonlinear resonance radius of a bubble is discussed as well as the analytical solution of the linearized equation of bubble pulsation. The mechanism of shock wave emission from a bubble into surrounding liquid is discussed. Inside a collapsing bubble, a shock wave is seldom formed due to lower temperature near the bubble wall. A liquid jet penetrates into a collapsing bubble near the solid surface. The bubble pulsation is influenced by the acoustic emissions from the surrounding bubbles, which is called *bubble–bubble interaction*. The origin of acoustic cavitation noise is discussed based upon results of numerical simulations. It is shown that surfactants and salts strongly retard bubble–bubble coalescence.

Keywords Rayleigh–Plesset equation · Keller equation · Rayleigh collapse Resonance radius · Shock wave · Jetting · Primary and secondary Bjerkens forces · Bubble–bubble interaction · Acoustic cavitation noise · Acoustic streaming

2.1 Rayleigh–Plesset Equation

A typical cavitation bubble is filled with vapor and non-condensable gas such as air. The pressure inside a bubble is higher than the liquid pressure at the bubble wall due to surface tension [1, 2]. The surface tension (σ) is the surface energy per unit

© The Author(s) 2018
K. Yasui, *Acoustic Cavitation and Bubble Dynamics*,
Ultrasound and Sonochemistry, https://doi.org/10.1007/978-3-319-68237-2_2

area and is 7.275×10^{-2} (N/m) (= J/m^2) for pure water at 20 °C. For a spherical bubble with a radius R, the surface energy is $4\pi\sigma R^2$ because the surface area is $4\pi R^2$. The work required to expand a bubble by dR in radius is $8\pi\sigma R$dR because the surface area becomes $4\pi(R + dR)^2 = 4\pi R^2 + 8\pi RdR$ [neglecting the $(dR)^2$ term]. Thus, the force needed to expand a bubble is $8\pi\sigma R$ because the work is the force multiplied by the distance moved (dR). The balance between the force inside and outside a bubble is expressed as $4\pi R^2 p_{in} = 4\pi R^2 p_B + 8\pi\sigma R$, where p_{in} is the pressure inside the bubble, and p_B is the liquid pressure at the bubble wall. Thus, the following relationship holds.

$$p_{in} = p_B + \frac{2\sigma}{R} \tag{2.1}$$

The second term on the right side of Eq. (2.1) is called the *Laplace pressure*. The pressure inside a bubble is higher than the liquid pressure at the bubble wall by the Laplace pressure. In Fig. 2.1, the Laplace pressure is shown as a function of bubble radius in pure water at 20 °C. The Laplace pressure is 1.5 bar for $R = 1$ μm and increases as the bubble radius decreases. For $R = 100$ nm (= 0.1 μm), it is as high as 15 bar (= 1.5×10^6 Pa $\approx$ 15 atm) [3].

Bubble dynamics such as violent bubble collapse is crudely described by the Rayleigh–Plesset equation. In its derivation, a spherical liquid volume with radius R_L surrounding a spherical bubble with radius R is considered with the center of a liquid volume at the center of a spherical bubble (Fig. 2.2) [2]. The radius of the liquid volume is much smaller than the wavelength of ultrasound in liquid; $R_L \ll \lambda$. When a bubble expands or collapses, the liquid volume also correspondingly expands or contracts, respectively. The kinetic energy of the liquid volume is estimated as follows: A spherical shell of liquid with thickness dr and radius r from the center of a spherical bubble has a kinetic energy of $1/2 \times 4\pi r^2 \rho_0 dr$ (the mass) $\times (dr/dt)^2$ (square of velocity), where ρ_0 is the equilibrium density of the liquid. The total

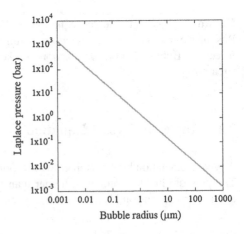

Fig. 2.1 Laplace pressure in pure water at 20 °C as a function of bubble radius [3]

Fig. 2.2 Derivation of the Rayleigh–Plesset equation of bubble pulsation. Reprinted with permission from Yasui [2]. Copyright (2015), Elsevier

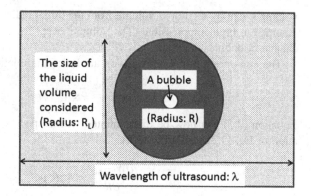

kinetic energy (E_K) of the liquid volume is the integration of the above quantity with respect to radius r from R to R_L, where R is the instantaneous bubble radius.

$$E_K = \frac{1}{2}\rho_0 \int_R^{R_L} \left(\frac{dr}{dt}\right)^2 4\pi r^2 dr = 2\pi\rho_0 R^3 \left(\frac{dR}{dt}\right)^2 \qquad (2.2)$$

where the liquid is assumed to be incompressible $\left(4\pi r^2 \frac{dr}{dt} = 4\pi R^2 \frac{dr}{dt}\right)$, and $R \ll R_L$.

When a bubble expands, it does work on the surrounding liquid. When a bubble collapses, the surrounding liquid does work on a bubble. In other words, a bubble does negative work on the surrounding liquid. The work (W_{bubble}) done by a bubble to the surrounding liquid can be expressed as follows:

$$W_{\text{bubble}} = \int_{R_0}^R 4\pi r^2 p_B dr \qquad (2.3)$$

where R_0 is the initial ambient bubble radius which is defined as the bubble radius in the absence of driving acoustic wave (ultrasound).

When a bubble expands, the liquid volume also expands. In other words, the liquid volume does work to the surrounding liquid. When a bubble collapses, the liquid volume contracts and does negative work on the surrounding liquid. The work (W_{liquid}) done by the liquid volume is expressed as follows.

$$W_{\text{liquid}} = p_\infty \Delta V = p_\infty \int_{R_0}^R 4\pi r^2 dr \qquad (2.4)$$

where p_∞ is the pressure at the surface of the liquid volume which is assumed to be the ambient static pressure plus the instantaneous acoustic pressure. ΔV is the volume swept by the liquid volume from the initial radius of R_L. Here, the liquid is assumed to be incompressible (The volume swept by a liquid is equivalent to the change in bubble volume.)

The conservation of energy yields the following relationship.

$$W_{\text{bubble}} = E_K + W_{\text{liquid}} \tag{2.5}$$

Equation (2.5) can now be differentiated with respect to R. Firstly, the differentiation of Eq. (2.3) yields Eq. (2.6).

$$\frac{\partial W_{\text{bubble}}}{\partial R} = 4\pi R^2 p_B \tag{2.6}$$

Secondly, the differentiation of Eq. (2.2) yields Eq. (2.7).

$$\frac{\partial E_K}{\partial R} = 6\pi \rho_0 R^2 \left(\frac{dR}{dt}\right)^2 + 4\pi \rho_0 R^3 \frac{d^2R}{dt^2} \tag{2.7}$$

where the following relationship has been used.

$$\frac{\partial}{\partial R}\left[\left(\frac{dR}{dt}\right)^2\right] = \frac{\partial(\dot{R}^2)}{\partial R} = \frac{1}{\dot{R}}\frac{\partial(\dot{R}^2)}{\partial t} = 2\ddot{R} = 2\frac{d^2R}{dt^2} \tag{2.8}$$

where "dot" denotes the time derivative (d/dt). Finally, the differentiation of Eq. (2.4) yields Eq. (2.9).

$$\frac{\partial W_{\text{liquid}}}{\partial R} = 4\pi R^2 p_\infty \tag{2.9}$$

From Eqs. (2.6), (2.7), and (2.9), differentiation of Eq. (2.5) with respect to R becomes Eq. (2.10).

$$\frac{p_B - p_\infty}{\rho_0} = \frac{3}{2}\dot{R}^2 + R\ddot{R} \tag{2.10}$$

When the bubble wall is moving, there is an additional term in Eq. (2.1) due to viscosity.

$$p_B = p_g + p_v - \frac{2\sigma}{R} - \frac{4\mu\dot{R}}{R} \tag{2.11}$$

where p_g and p_v are partial pressures of non-condensable gas and vapor, respectively, $(p_{\text{in}} = p_g + p_v)$, and μ is the liquid viscosity. In the derivation of the last term on right-hand side of Eq. (2.11), the incompressibility of liquid is assumed to be $4\pi r^2\dot{r} = 4\pi R^2\dot{R}$ because the term is derived from $2\mu\frac{\partial \dot{r}}{\partial r}\big|_{r=R}$.

Finally, the Rayleigh–Plesset equation is derived by inserting Eq. (2.11) into Eq. (2.10).

$$R\ddot{R} + \frac{3}{2}\dot{R}^2 = \frac{1}{\rho_0}\left[p_g + p_v - \frac{2\sigma}{R} - \frac{4\mu\dot{R}}{R} - p_0 - p_s(t)\right] \tag{2.12}$$

where p_0 is the ambient static pressure, and $p_s(t)$ is the instantaneous acoustic pressure at time t $(p_\infty = p_0 + p_s(t))$. As the incompressibility of the liquid is assumed in the derivation of Eq. (2.12), the equation is no longer valid when a bubble violently collapses with speed comparable to sound velocity in the liquid.

2.2 Rayleigh Collapse

After bubble expansion during the rarefaction phase of ultrasound, a bubble violently collapses if the ambient bubble radius reaches a critical range. The range of ambient radius for an active bubble is discussed in Sect. 2.12. In this section, the cause for the violent collapse of a bubble is discussed using the Rayleigh–Plesset equation [2]. From Eq. (2.12), the bubble wall acceleration $(\ddot{R})$ is expressed as follows.

$$\ddot{R} = -\frac{3\dot{R}^2}{2R} + \frac{1}{\rho_0 R}\left[p_g + p_v - \frac{2\sigma}{R} - \frac{4\mu\dot{R}}{R} - p_0 - p_s(t)\right] \tag{2.13}$$

When a bubble violently collapses and $\dot{R}^2$ increases, the first term on right-hand side of Eq. (2.13) becomes dominant, and the second term becomes negligible. In this situation, the following relationship nearly holds.

$$\ddot{R} \approx -\frac{3\dot{R}^2}{2R} \tag{2.14}$$

This means that the bubble wall acceleration $(\ddot{R})$ is always negative. It results in a decrease in the bubble wall velocity $(\dot{R})$. As the bubble wall velocity is negative during the bubble collapse, the magnitude of bubble wall velocity increases with time. Then, the magnitude on the right-hand side of Eq. (2.14) further increases, and the magnitude of the bubble wall acceleration further increases. In this way, the bubble collapse freely accelerates, which is the reason for the violent bubble collapse called the *Rayleigh collapse*. Finally, the pressure (p_g) inside a bubble dramatically increases when the density inside a bubble becomes comparable to that of condensed phase (liquid). Then, the second term in Eq. (2.13) becomes dominant, and the bubble collapse stops as the bubble wall acceleration takes a large positive value.

Fig. 2.3 Spherically inward flow as the mechanism for the violent collapse of a cavitation bubble. Reprinted with permission from Yasui [2]. Copyright (2015), Elsevier

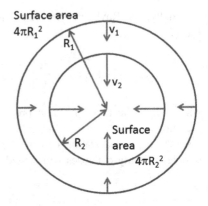

The question that one may ask is: What is the physical reason for the freely accelerating collapse of a bubble? There are two reasons for it. One is the inertia of the surrounding liquid which flows toward a bubble during the bubble collapse. Thus, cavitation with such violent collapse of bubbles is called *inertial cavitation* (or *transient cavitation*). The other reason for the freely accelerating collapse is the geometry of a spherical collapse. Due to the conservation of mass, the velocity of the liquid toward the center of a bubble increases as the distance from the center of a bubble decreases. Let us consider two concentric spherical surfaces in a liquid with their center at the center of a bubble (Fig. 2.3) [2]. The radii of two concentric spherical surfaces are R_1 and R_2 $(R_1 > R_2)$. The mass (liquid) flow rate is $4\pi R_1^2 v_1$ and $4\pi R_2^2 v_2$ at the spherical surface of radii R_1 and R_2, respectively. The conservation of mass yields $4\pi R_1^2 v_1 = 4\pi R_2^2 v_2$. Thus, the velocity for smaller radius is larger, $v_2 = \left(\frac{R_1}{R_2}\right)^2 v_1 > v_1$. This is a nature of a spherical geometry which causes freely accelerating collapse of a bubble.

2.3   Keller Equation

The effect of liquid compressibility is approximately taken into account in bubble dynamics equation as follows: The starting equations are the continuity equation (conservation of mass) (Eq. 2.15) and Euler equation (equation of motion) (Eq. 2.16) [4–6].

$$\frac{\partial \rho}{\partial t} + \nabla \cdot (\rho \vec{u}) = \frac{\partial \rho}{\partial t} + \vec{u} \cdot \nabla \rho + \rho \nabla \cdot \vec{u} = 0 \qquad (2.15)$$

$$\rho \frac{D\vec{u}}{Dt} = \rho \left(\frac{\partial \vec{u}}{\partial t} + (\vec{u} \cdot \nabla)\vec{u} \right) = -\nabla p \qquad (2.16)$$

where $\nabla = \left(\frac{\partial}{\partial x}, \frac{\partial}{\partial y}, \frac{\partial}{\partial z}\right)$, $\nabla \cdot (\rho \vec{u}) = \frac{\partial}{\partial x}(\rho u_x) + \frac{\partial}{\partial y}(\rho u_y) + \frac{\partial}{\partial z}(\rho u_z)$, ρ is the instantaneous local density of liquid, $\vec{u}$ is the instantaneous local velocity of liquid $(\vec{u} = (u_x, u_y, u_z))$, $\nabla \rho = \left(\frac{\partial \rho}{\partial x}, \frac{\partial \rho}{\partial y}, \frac{\partial \rho}{\partial z}\right)$, $\frac{D}{Dt}$ is the material time derivative $\left(\frac{D}{Dt} = \frac{\partial}{\partial t} + \vec{u} \cdot \nabla\right)$, $\vec{u} \cdot \nabla = u_x \frac{\partial}{\partial x} + u_y \frac{\partial}{\partial y} + u_z \frac{\partial}{\partial z}$, and p is the instantaneous local pressure of the liquid. In Eq. (2.16), the effects of gravitational force and liquid viscosity are neglected. The derivation of the above equations is described in detail in textbooks of fluid dynamics [7].

Here, it is assumed that the velocity field of the liquid around a pulsating bubble has only a radial component. In this case, the liquid flow is irrotational, and the velocity field is expressed by using a velocity potential (ϕ).

$$\vec{u} = \nabla \phi = \frac{\partial \phi}{\partial r} \vec{e_r} \tag{2.17}$$

where r is radial distance from the center of a bubble, and $\vec{e_r}$ is a radial unit vector. Then, Eqs. (2.15) and (2.16) are expressed as Eqs. (2.18) and (2.19), respectively.

$$\frac{\partial \rho}{\partial t} + \left(\frac{\partial \phi}{\partial r}\right)\left(\frac{\partial \rho}{\partial r}\right) + \rho \Delta \phi = 0 \tag{2.18}$$

$$\rho \left[\frac{\partial^2 \phi}{\partial t \partial r} + \left(\frac{\partial \phi}{\partial r}\right)\left(\frac{\partial^2 \phi}{\partial r^2}\right)\right] = -\frac{\partial p}{\partial r} \tag{2.19}$$

From Eqs. (2.18) and (2.19), the following modified wave equation is derived (for a detailed method of derivation, see Ref. [4]).

$$\Delta \phi - \frac{1}{c^2}\frac{\partial^2 \phi}{\partial t^2} = \frac{1}{c^2}\left(\frac{\partial \phi}{\partial r}\right)\left(\frac{\partial^2 \phi}{\partial r \partial t}\right) - \frac{1}{\rho}\left(\frac{\partial \rho}{\partial r}\right)\left(\frac{\partial \phi}{\partial r}\right) \tag{2.20}$$

where c is the instantaneous local sound velocity. In the derivation of the Keller equation of bubble dynamics, the right-hand side of Eq. (2.20) is neglected and the wave equation (Eq. 2.21) can be used. Thus, the Keller equation is an approximate equation which is only valid when $\frac{|\dot{R}|}{c_\infty} \ll 1$, where c_∞ is the sound velocity at ambient condition.

$$\Delta \phi - \frac{1}{c_\infty^2}\frac{\partial^2 \phi}{\partial t^2} = 0 \tag{2.21}$$

By integrating Eq. (2.19) with respect to r, the following approximate equation may be derived [4].

$$\frac{\partial \phi}{\partial t} + \frac{1}{2}u^2 + \frac{p - p_\infty}{\rho_{L,\infty}} = 0 \tag{2.22}$$

where p_∞ is the ambient pressure, and the liquid density is assumed constant at ambient conditions $\left(\rho = \rho_{L,\infty} = \text{const.}\right)$. The boundary condition is given as follows.

$$\left(\frac{\partial \phi}{\partial r}\right)_{r=R} = \dot{R} \tag{2.23}$$

The general solution of the wave equation (Eq. 2.21) under spherical symmetry is given as follows.

$$\phi = -\frac{f\left(t - \frac{r}{c_\infty}\right)}{r} - \frac{g\left(t + \frac{r}{c_\infty}\right)}{r} \tag{2.24}$$

where f and g are arbitrary functions. From Eqs. (2.23) and (2.24), Eq. (2.25) is obtained.

$$\frac{f'}{R} = c_\infty \left[\dot{R} + \frac{\phi(R)}{R}\right] + \frac{g'}{R} \tag{2.25}$$

where $'$ means derivative. From Eqs. (2.24) and (2.25), Eq. (2.26) is obtained.

$$\left(\frac{\partial \phi}{\partial t}\right)_{r=R} = -c_\infty \left[\dot{R} + \frac{\phi(R)}{R}\right] - \frac{2g'}{R} \tag{2.26}$$

Inserting Eqs. (2.23) and (2.26) into Eq. (2.22) yields Eq. (2.27).

$$-c_\infty \left[\dot{R} + \frac{\phi(R)}{R}\right] - \frac{2g'}{R} + \frac{1}{2}\dot{R}^2 + \frac{p_B - p_\infty}{\rho_{L,\infty}} = 0 \tag{2.27}$$

where p_B is the liquid pressure at the bubble. Thus, multiplying Eq. (2.27) by R and differentiating by t yields Eq. (2.28).

$$0 = -c_\infty \left[\dot{R}^2 + R\ddot{R} + \left(\frac{d\phi}{dt}\right)_{r=R}\right]$$
$$- 2\frac{dg'}{dt} + \frac{1}{2}\dot{R}^3 + R\dot{R}\ddot{R} + \frac{R}{\rho_{L,\infty}}\frac{dp_B}{dt} + \frac{1}{\rho_{L,\infty}}\dot{R}(p_B - p_\infty) \tag{2.28}$$

where $''$ means the second derivative. From Eqs. (2.22) and (2.23), Eq. (2.29) is obtained.

$$\left(\frac{d\phi}{dt}\right)_{r=R} = \frac{\partial\phi}{\partial t} + \frac{\partial\phi}{\partial r}\frac{dr}{dt} = -\frac{1}{2}\dot{R}^2 - \frac{p_B - p_\infty}{\rho_{L,\infty}} + \dot{R}^2 \tag{2.29}$$

When the incident field is a plane acoustic wave with an angular frequency ω and a pressure amplitude A, the following relationship holds [4].

$$2g'' = -\frac{c_\infty}{\rho_{L,\infty}} A \sin\omega t \tag{2.30}$$

Inserting Eqs. (2.29) and (2.30) into Eq. (2.28) yields the equation of bubble dynamics called the Keller equation (Eq. 2.31).

$$\left(1 - \frac{\dot{R}}{c_\infty}\right) R\ddot{R} + \frac{3}{2}\dot{R}^2\left(1 - \frac{\dot{R}}{3c_\infty}\right) = \frac{1}{\rho_{L,\infty}}\left(1 + \frac{\dot{R}}{c_\infty}\right)[p_B - p_s(t) - p_\infty] + \frac{R}{c_\infty\rho_{L,\infty}}\frac{dp_B}{dt} \tag{2.31}$$

where $A \sin\omega t$ is replaced by $p_s(t)$.

As already noted, the Keller equation is valid only when $\frac{|\dot{R}|}{c_\infty} \ll 1$. However, in numerical simulations using the Keller equation, this condition is often violated during a violent bubble collapse. When $|\dot{R}|$ exceeds c_∞, $\left(1 + \frac{\dot{R}}{c_\infty}\right)$ in the right-hand side of Eq. (2.31) changes sign, and the error in numerical calculations becomes significant. To avoid this error, c_∞ in Eq. (2.31) is sometimes replaced by the sound velocity in the liquid at the bubble wall ($c_{L,B}$). It dramatically increases as the liquid pressure at the bubble wall increases as follows [8, 9].

$$c_{L,B} = \sqrt{\frac{7.15(p_B + B)}{\rho_{L,i}}} \tag{2.32}$$

where $c_{L,B}$ is the sound velocity in the liquid water at the bubble wall, $B = 3.049 \times 10^8$ (Pa), and $\rho_{L,i}$ is the liquid density at the bubble wall. The liquid density at the bubble wall is a function of pressure and temperature at the bubble wall [6]. In numerical simulations using the Keller equation, the bubble wall speed $|\dot{R}|$ sometimes still exceeds $c_{L,B}$. In that case, the bubble wall speed is replaced by $c_{L,B}$ because the bubble wall speed never exceeds $c_{L,B}$ according to the following arguments [9].

For steady flows $\left(\frac{\partial u}{\partial t} = 0\right)$, the Euler equation (Eq. 2.16) yields Eq. (2.33) [10].

$$udu = -\frac{dp}{\rho} = -\frac{dp}{d\rho}\frac{d\rho}{\rho} = -c^2\frac{d\rho}{\rho} \tag{2.33}$$

where the following relationship for sound velocity is used.

$$c = \sqrt{\frac{\mathrm{d}p}{\mathrm{d}\rho}} \tag{2.34}$$

Using Mach number $M = \frac{u}{c}$, Eq. (2.33) becomes Eq. (2.35).

$$\frac{\mathrm{d}\rho}{\rho} = -M^2 \frac{\mathrm{d}u}{u} \tag{2.35}$$

For radial steady flows $\left(\frac{\partial \rho}{\partial t} = 0\right)$ toward the center of a bubble, the conservation of mass requires the following relationship [10].

$$\rho u A_s = \text{independent of } r \tag{2.36}$$

where A_s is the surface area of a sphere $(A_s = 4\pi r^2)$. The differentiation of Eq. (2.36) yields Eq. (2.37).

$$uA_s \cdot \mathrm{d}\rho + \rho A_s \cdot \mathrm{d}u + \rho u \cdot \mathrm{d}A_s = 0 \tag{2.37}$$

Equation (2.37) is equivalent to the following equation.

$$\frac{\mathrm{d}\rho}{\rho} + \frac{\mathrm{d}u}{u} + \frac{\mathrm{d}A_s}{A_s} = 0 \tag{2.38}$$

Inserting Eq. (2.35) into Eq. (2.38) yields Eq. (2.39).

$$\frac{\mathrm{d}u}{u} = -\frac{\frac{\mathrm{d}A_s}{A_s}}{(1 - M^2)} = -\frac{2\mathrm{d}r/r}{(1 - M^2)} \tag{2.39}$$

where $A_s = 4\pi r^2$ is used. As already discussed in Sect. 2.2, the magnitude of the liquid velocity increases (the liquid velocity decreases because $u < 0$ for inward liquid flow) as the radial distance from the center of the bubble decreases ($\mathrm{d}u < 0$ for $\mathrm{d}r < 0$). From Eq. (2.39), it implies that $|M| < 1$.

In the above discussion, a liquid flow is assumed as steady ($\frac{\partial u}{\partial t} = 0$ and $\frac{\partial \rho}{\partial t} = 0$). Under typical conditions of a bubble collapse, the terms $\frac{\partial u}{\partial t}$ and $\frac{\partial \rho}{\partial t}$ are actually negligible compared to the other terms in the Euler equation and the equation of continuity, respectively. However, further studies are required on the upper limit of the bubble wall speed in the case of non-steady flows. Furthermore, the method of numerical simulations of the Keller equation discussed above is rather "tricky." Rigorous derivation of more accurate equation of the bubble dynamics is an important focus and task. Of relevance, there have been a few studies based upon the direct numerical simulations of the bubble collapse employing fundamental equations of fluid dynamics [11, 12].

2.4 Method of Numerical Simulations

For quantitative discussions on bubble pulsation, numerical simulations of the Keller equation and other equations of bubble dynamics are required. The simplest method of numerical simulation is the Euler method [2, 13].

$$R(t + \Delta t) = R(t) + \dot{R}(t)\Delta t \tag{2.40}$$

where $R(t)$ is the instantaneous bubble radius at time t, Δt is a time step in numerical simulation. In numerical simulations, the continuous time is divided into a large number of discrete times with a small unit step (Δt). Equation (2.40) is derived directly from the definition of the time derivative.

$$\dot{R}(t) = \lim_{\Delta t \to 0} \frac{R(t + \Delta t) - R(t)}{\Delta t} \tag{2.41}$$

The bubble wall velocity ($\dot{R}$) is also calculated in the same manner.

$$\dot{R}(t + \Delta t) = \dot{R}(t) + \ddot{R}(t)\Delta t \tag{2.42}$$

The bubble wall acceleration ($\ddot{R}(t)$) is calculated by the Keller equation (Eq. 2.31) and other equations of bubble dynamics. For numerical simulations, initial values of R and $\dot{R}$ are required. When the Keller equation is used, initial values of p_B and $\frac{dp_B}{dt}$ are also required. At an arbitrary time, the pressure inside a bubble ($p_{in} = p_g + p_v$) needs to be calculated in order to calculate p_B and $\ddot{R}(t)$. For this purpose, the van der Waals equation of state is used [14].

$$\left[p_{in} + \frac{a_v}{v^2} \right] (v - b_v) = R_g T \tag{2.43}$$

where a_v and b_v are the van der Waals constants, v is the molar volume, R_g is the gas constant, and T is the temperature inside a bubble. The molar volume v is calculated as follows.

$$v = \frac{4\pi R^3}{3} \cdot \frac{N_A}{n_t} \tag{2.44}$$

where N_A is the Avogadro number ($= 6.02 \times 10^{23}$ mol^{-1}), and n_t is the total number of molecules inside a bubble. In order to calculate the pressure inside a bubble (p_{in}), the temperature and the total number of molecules inside a bubble are required. The temperature (T) inside a bubble is approximately calculated from internal thermal energy (E) of a bubble as follows [14].

$$E = \frac{T}{N_A} \sum_\alpha n_\alpha C_{V,\alpha} - \left(\frac{n_t}{N_A}\right)^2 \frac{a_v}{V} \qquad (2.45)$$

where n_α is the number of molecules of species α inside a bubble, $C_{V,\alpha}$ is the molar heat capacity at constant volume of species α, the summation is for all the gas and vapor species inside a bubble, and V is the bubble volume $\left(V = \frac{4}{3}\pi R^3\right)$. The derivation of Eq. (2.45) is as follows [15, 16]. The internal energy (E) of a bubble is a function of temperature (T) and volume (V) of a bubble for the van der Waals gas.

$$dE = \left(\frac{\partial E}{\partial T}\right)_V dT + \left(\frac{\partial E}{\partial V}\right)_T dV \qquad (2.46)$$

From the definition of the molar heat capacity at constant volume, the following relationship holds.

$$\left(\frac{\partial E}{\partial T}\right)_V = \frac{1}{N_A} \sum_\alpha n_\alpha C_{V,\alpha} \qquad (2.47)$$

For the van der Waals gas, the second term on the right-hand side of Eq. (2.46) is nonzero.

$$\left(\frac{\partial E}{\partial V}\right)_T = T\left(\frac{\partial p_{in}}{\partial T}\right)_V - p_{in} = \left(\frac{n_t}{N_A}\right)^2 \frac{a_v}{V^2} \qquad (2.48)$$

Using Eqs. (2.47) and (2.48), integration of Eq. (2.46) yields Eq. (2.45) assuming temperature-independent molar heat at constant volume. The temporal change in the total number (n_t) of molecules as well as those of species α inside a bubble is discussed in Sects. 2.5–2.9.

The temporal change (ΔE) in an internal thermal energy of a bubble is calculated as follows [14].

$$\Delta E = -p_{in}\Delta V + 4\pi R^2 \dot{m} e_{H_2O} \Delta t + 4\pi R^2 \kappa \frac{\partial T}{\partial r}\bigg|_{r=R} \Delta t + \frac{4}{3}\pi R^3 \Delta t \sum_r \left(r_{yb} - r_{yf}\right)\Delta H_{yf}$$

$$+ \sum_{\alpha'} e_{\alpha'}\Delta n_{\alpha'} + \left[-\frac{3}{5}M_{in}R\ddot{R}\right]\Delta t$$

$$(2.49)$$

where $\dot{m}$ is the rate of non-equilibrium evaporation at the bubble wall, e_{H_2O} is the energy carried by an evaporating or condensing vapor molecule, κ is the thermal conductivity of a mixture of gases and vapor, $\frac{\partial T}{\partial r}\big|_{r=R}$ is the temperature gradient inside a bubble at the bubble wall, $r_{\gamma b}$ and $r_{\gamma f}$ are the backward and forward reaction rates, respectively, of chemical reaction γ per unit volume and unit time, $\Delta H_{\gamma f}$ is the enthalpy change in the forward reaction (when $\Delta H_{\gamma f} < 0$, i.e., the reaction is exothermic), the summation is for all the chemical reactions occurring inside a bubble, $e_{\alpha'}$ is the energy carried by a diffusing gas molecule of species α', $\Delta n_{\alpha'}$ is the change in number of molecules of species α' by diffusion, the summation is for all the gas species except vapor inside a bubble, and M_{in} is the total mass of gases and vapor inside a bubble. The first term on the right-hand side of Eq. (2.49) is the pV work done by the surrounding liquid on a bubble. The second term is the energy change associated with evaporation or condensation. The third term is the energy change due to the thermal conduction. The fourth term is the heat of chemical reactions. The fifth term is the energy change due to diffusion. The last term is included only when the quantity in the brackets is positive and is the heat due to the decrease in kinetic energy of gases and vapor inside a collapsing bubble. The derivation of the last term is given in Ref. [14]. More details of the model are described in Refs. [14, 17], and there are other similar models from various researchers [18, 19].

The results of numerical simulations based upon the present model of the bubble dynamics are shown in Figs. 2.4 and 2.5 under a condition of single-bubble sonoluminescence (SBSL) [2]. A bubble expands during the rarefaction phase of ultrasound (Fig. 2.4a). In order to make a bubble initially expand, $p_s(t)$ in Eq. (2.31) is assumed as $p_s(t) = -A \sin \omega t$. During the compression phase of ultrasound, a bubble violently collapses followed by bouncing motion (weaker

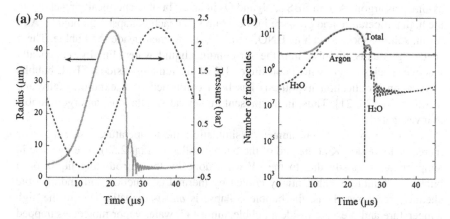

Fig. 2.4 Results of numerical simulations of bubble pulsation under a SBSL condition as a function of time for one acoustic cycle. The frequency and pressure amplitude of the acting ultrasound are 22 kHz and 1.32 bar, respectively. The ambient (equilibrium) bubble radius is 4 μm for an argon (Ar) bubble in 20 °C water. **a** The bubble radius (solid line) and the pressure $[p_\infty + p_s(t)]$ (dotted line) **b** The number of molecules inside a bubble on a logarithmic scale. Reprinted with permission from Yasui [2]. Copyright (2015), Elsevier

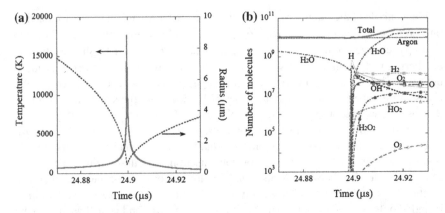

Fig. 2.5 Results of numerical simulations around the bubble collapse as a function of time, specifically for 0.06 μs (the condition is the same as that of Fig. 2.4). **a** The bubble radius (dotted line) and the temperature inside a bubble (solid line). **b** The number of molecules inside a bubble on a logarithmic scale. At $t = 24.93$ ms (the right end of the graph), the main chemical products are H_2 (1×10^8 in number of molecules), O_2 (4×10^7), O (3×10^7), H (2×10^7), H_2O_2 (1×10^7), and OH (7×10^6). Reprinted with permission from Yasui [2]. Copyright (2015), Elsevier

pulsation). During the bubble expansion, the number of H_2O molecules inside a bubble increases by evaporation at the bubble wall as the pressure inside a bubble decreases. During the bubble collapse, the number of H_2O molecules decreases by condensation at the bubble wall.

In the present numerical simulations, the non-condensable gas inside a bubble is assumed as argon (Ar). In SBSL, N_2 and O_2 in an air bubble chemically react due to the high temperature and pressure inside a bubble at each collapse of a bubble. As a result, soluble species such as HNO_x and NO_x are formed inside the bubble. These species gradually dissolve into the surrounding liquid water. Finally, chemically inactive species argon, which constitutes 1% of air, remains inside a SBSL bubble. This argon rectification hypothesis has been confirmed both experimentally and theoretically [20, 21]. Thus, in the present numerical simulations, an argon bubble is investigated.

According to the present numerical simulation, the temperature inside a bubble increases to 18,000 K at the end of the bubble collapse (Fig. 2.5a). The increase in temperature is mostly due to the pV work done by the surrounding liquid on a bubble. A bubble is substantially cooled by thermal conduction and endothermic chemical reactions. Thus, the bubble collapse is *quasi*-adiabatic. Due to the high temperature and pressure inside a bubble, almost all water vapor molecules trapped inside a bubble are dissociated, and hydrogen (H_2), oxygen (O_2), and hydroxyl radicals ($OH^\bullet$) are created (Fig. 2.5b). OH radicals play an important role in sonochemistry as previously discussed.

2.5 Non-equilibrium Evaporation and Condensation

There are two types of mass transfer across the bubble wall. One is non-equilibrium evaporation and condensation of (water) vapor at the bubble wall. The other is the diffusion of non-condensable gases across the bubble wall. In this section, non-equilibrium evaporation and condensation of water vapor is discussed. There are two steps in non-equilibrium evaporation and condensation processes (Fig. 2.6) [22]. One is the diffusion of water vapor inside a bubble. The other is the phase change at the bubble wall. According to full numerical simulations by Storey and Szeri [23], the diffusion of water vapor inside a bubble is sometimes the rate-determining step. According to their full numerical simulations [23], the molar fraction of water vapor near the bubble wall inside a bubble is about one order of magnitude smaller than that at the center of a bubble near the final stage of a violent bubble collapse. However, in their numerical simulations [23], fluid velocity inside a collapsing bubble is assumed to have only radial component. The possible appearance of non-radial component of fluid velocity inside a collapsing bubble such as in turbulence should be studied in future. If turbulence occurs inside a collapsing bubble, molar fraction of water vapor is more homogeneous.

There is another issue on inhomogeneous molar fraction of water vapor inside a collapsing bubble. According to numerical simulations by Storey and Szeri [8], there are intense temperature and pressure gradients inside a collapsing bubble. These gradients drive relative mass diffusion which overwhelms diffusion driven by concentration gradients. These thermal and pressure diffusion processes result in a robust compositional inhomogeneity in the bubble which lasts for several orders of magnitude longer than the temperature peak. In their study on the mixture segregation [8], a mixture of He and Ar gases was investigated. Mixture segregation occurs for a mixture of gases with large difference in their molecular weights. When molecular weight of non-condensable gas is largely different from that of water vapor, water vapor and non-condensable gas are expected to be mildly segregated inside a collapsing bubble [8, 24, 25]. In this case, rate of non-equilibrium condensation during bubble collapse is strongly influenced by mixture segregation. Further studies are required on whether mixture segregation occurs inside a collapsing bubble.

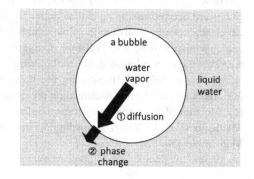

Fig. 2.6 Two steps in vapor transport inside a bubble to the bubble wall. Reprinted with permission from Yasui et al. [22]. Copyright (2004), Taylor & Francis

If the rate-determining step is the phase change, the rate ($\dot{m}$) of non-equilibrium evaporation and condensation at the bubble wall is given by the following equations [14, 26, 27].

$$\dot{m} = \dot{m}_{eva} - \dot{m}_{con} \tag{2.50}$$

$$\dot{m}_{eva} = \frac{10^3 N_A}{M_{H_2O}} \frac{\alpha_M}{\sqrt{2\pi R_v}} \frac{p_v^*}{\sqrt{T_{L,i}}} \tag{2.51}$$

$$\dot{m}_{con} = \frac{10^3 N_A}{M_{H_2O}} \frac{\alpha_M}{\sqrt{2\pi R_v}} \frac{\Gamma p_v}{\sqrt{T_B}} \tag{2.52}$$

where $\dot{m}_{eva}$ and $\dot{m}_{con}$ are actual rates of evaporation and condensation ($\dot{m}$ is the net rate of evaporation), M_{H_2O} is the molecular weight of H_2O (=18 g/mol), and α_M is the accommodation coefficient for evaporation or condensation. R_v is the gas constant of water vapor in (J/kg K), p_v^* is the saturated vapor pressure at the liquid temperature ($T_{L,i}$) at the bubble wall, p_v is actual vapor pressure inside a bubble, T_B is the temperature at the bubble wall inside a bubble, and the correction factor (Γ) is given as follows.

$$\Gamma = e^{-\Omega^2} - \Omega\sqrt{\pi}\left(1 - \frac{2}{\sqrt{\pi}} \int_0^\Omega e^{-x^2} dx\right) \tag{2.53}$$

where

$$\Omega = \frac{\dot{m}}{p_v}\left(\frac{R_v T}{2}\right)^{1/2} \tag{2.54}$$

The actual vapor pressure (p_v) inside a bubble is given as follows.

$$p_v = \frac{n_{H_2O}}{n_t} p_{in} \tag{2.55}$$

where n_{H_2O} is the number of H_2O molecules inside a bubble. Equations (2.50)–(2.52) are derived assuming a thin boundary layer near the liquid–gas interface in which the velocity distribution of molecules is Maxwell–Boltzmann [26]. From the velocity distribution, the collision frequency of molecules at the surface of a boundary layer is calculated. By multiplying it with the accommodation coefficient which is a probability of escaping the boundary layer for a molecule, the actual rates of evaporation and condensation are obtained. The accommodation coefficient is a function of the liquid temperature at the bubble wall: It decreases from 0.35 at 350 K to 0.05 at 500 K according to the molecular dynamics simulations by Matsumoto [14, 28].

According to numerical simulations, the non-equilibrium effect is only dominant at a bubble collapse [29]. At a strong collapse, the vapor pressure (p_v) inside a bubble is higher than the saturated vapor pressure (p_v^*) at the liquid temperature at the bubble wall $(T_{L,i})$ by more than one order of magnitude. On the other hand, during the bubble expansion, the vapor pressure is nearly identical to the saturated vapor pressure (nearly in equilibrium).

When diffusion is sometimes rate-determining step, the reader should refer to Refs. [18, 19].

2.6 Liquid Temperature at the Bubble Wall

Another important problem is the liquid temperature at the bubble wall. In numerical simulations, it is sometimes assumed to be identical to the ambient liquid temperature [18, 19]. However, there are some experimental evidences on the substantial increase in liquid temperature at the bubble wall [30–32]. For example, Suslick et al. [30] experimentally studied reaction rates of metal carbonyls in alkane solvent as a function of solvent vapor pressure when solutions were irradiated with ultrasound at 20 kHz. From the vapor pressure-independent component, they estimated the temperature at the interface between a bubble and liquid as ~1900 K. Hua et al. [31] experimentally studied sonochemical degradation of nonvolatile hydrophobic p-nitrophenyl acetate and concluded that the degradation was accelerated by supercritical water formed at the bubble interface region (Fig. 2.7) [33]. Supercritical water is defined as water at temperature and pressure higher than the critical ones (647 K and 221 bar, respectively). Moriwaki et al. [32] experimentally studied the sonochemical degradation of anionic surfactants and concluded that they were pyrolyzed at the interfacial region of a bubble. Thus, for accurate numerical simulations, the increase in temperature at the interface region of a bubble should be taken into account. There are some models available to determine the liquid temperature $(T_{L,i})$ at the bubble wall [6, 8, 34–36]. According

Fig. 2.7 Three regions for a cavitation bubble. Reprinted with permission from Yasui [33]. Copyright (2016), Springer

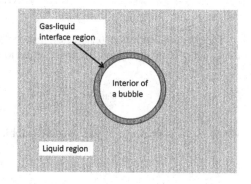

to the full numerical simulations by Storey and Szeri [23], the temperature of the interface exceeds the critical point (647 K) for about 2 ns at around the minimum bubble radius. The thickness of supercritical region is only about 10 nm. More studies are required to further elucidate the interface temperature at the bubble collapse.

2.7 Gas Diffusion (Rectified Diffusion)

Gas diffusion across the bubble surface is a complex phenomenon when a bubble is pulsating under ultrasound because the gas concentration in the liquid adjacent to bubble wall changes in an intricate way. Eller and Flynn [37] solved this problem in 1965 using material coordinates, which move as the liquid element moves. They solved the diffusion equation for a pulsating bubble by time averaging over an acoustic period as follows [37].

$$\left\langle \frac{dn_{gas}}{dt} \right\rangle = 4\pi R_0 D_{gas} \left[A_R + R_0 \left(\frac{B_R}{\pi D_{gas} t} \right)^{1/2} \right] c_{i0} \left(\frac{c_\infty}{c_{i0}} - \frac{A_R}{B_R} \right) \tag{2.56}$$

where n_{gas} is the number of molecules of non-condensable gas inside a bubble, $\langle \rangle$ means time-averaged value, R_0 is the ambient bubble radius, D_{gas} is the diffusion coefficient of the gas in the liquid, c_{i0} is the gas concentration at the bubble wall in the liquid at ambient bubble radius (R_0), c_∞ is the ambient gas concentration in the liquid, and A_R and B_R is defined as follows.

$$A_R = \frac{1}{T_a} \int_0^{T_a} \left(\frac{R}{R_0} \right) dt \tag{2.57}$$

$$B_R = \frac{1}{T_a} \int_0^{T_a} \left(\frac{R}{R_0} \right)^4 dt \tag{2.58}$$

The gas concentration (c_{i0}) at the bubble wall in the liquid at ambient bubble radius (R_0) is given by the following equation.

$$c_{i0} = c_\infty \frac{p_{gas,0}}{p_\infty} \approx c_\infty \left(1 + \frac{2\sigma}{R_0 p_\infty} \right) \tag{2.59}$$

where $p_{gas,0}$ is pressure of gas inside a bubble at ambient bubble radius.

From the time-averaged rate of gas diffusion in Eq. (2.56), the instantaneous rate of gas diffusion is postulated in Ref. [38] as follows.

$$\frac{dn_{gas}}{dt} = 4\pi R^2 D_{gas} \frac{A_R}{B_R} \frac{c_\infty - c_i}{(R_0/R)^2 R_0} \tag{2.60}$$

where c_i is the instantaneous gas concentration at the bubble wall in the liquid. In the derivation of Eq. (2.60) from Eq. (2.56), the second term in the square brackets in Eq. (2.56) is omitted, and the bubble pulsation is assumed to be isothermal as follows.

$$c_i = c_\infty \frac{p_{gas}}{p_\infty} \approx c_\infty \left(1 + \frac{2\sigma}{R_0 p_\infty}\right)\left(\frac{R_0}{R}\right)^3 \tag{2.61}$$

where p_{gas} is the instantaneous gas pressure inside a bubble, and $p_{gas}\left(\frac{4}{3}\pi R^3\right) = p_{gas,0}\left(\frac{4}{3}\pi R_0^3\right)$ is used. This assumption is justified as the bubble expansion is nearly isothermal and gas diffusion occurs mostly during bubble expansion.

Gas diffuses into a bubble during bubble expansion as the pressure inside a bubble is lower than the ambient pressure [$c_\infty > c_i$ in Eq. (2.60)]. During bubble collapse, the gas diffuses out of a bubble into surrounding liquid as the pressure inside a bubble is higher than the ambient pressure [$c_\infty < c_i$ in Eq. (2.60)]. When a bubble is strongly pulsating, the gas diffusion into a bubble during the bubble expansion overwhelms that out of a bubble during the bubble collapse; this process is called *rectified diffusion* [39, 40]. There are two main reasons in rectified diffusion. One is the larger surface area during the bubble expansion than that during the bubble collapse; this is called *area effect*. The other is larger magnitude of gradient of gas concentration during the bubble expansion than that during the bubble collapse. This is due to thinner material (liquid) element during the bubble expansion compared to that during the bubble collapse; this is called *shell effect* (Fig. 2.8). In Eq. (2.60), the shell effect is expressed by $\left(\frac{R_0}{R}\right)^2$ in the denominator. The area effect is expressed by $4\pi R^2$ in Eq. (2.60).

Growth rate of a pulsating bubble by rectified diffusion is defined as the rate of increase in ambient bubble radius. The growth rate by rectified diffusion strongly

Fig. 2.8 Shell effect in rectified diffusion

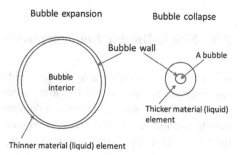

depends upon the operating conditions, such as the acoustic pressure amplitude and the acoustic frequency as well as the surface tension. The growth rate of a bubble with an initial radius of 35 μm was experimentally measured as a few micrometers per 100 s when the ultrasonic frequency and pressure amplitude were 22.1 kHz and 0.3 bar, respectively, in air-saturated water [41]. At acoustic pressure amplitude of 2 bar at 26.5 kHz, growth rate by rectified diffusion is numerically calculated to range from 10 to several 100 μm/s depending upon the initial ambient radius in nearly gas-saturated water [42].

2.8 Chemical Kinetic Model

Generally speaking, chemical reactions inside a collapsing bubble are in non-equilibrium [43]. Thus, it is necessary to use chemical kinetic model for numerical simulations of chemical reactions inside a bubble. For example, the rate of the following reaction is given by Eq. (2.63) [14, 35].

$$H_2O + M \rightarrow H + OH + M \tag{2.62}$$

$$r_f = A_f T^{b_f} e^{-C_f/T} [H_2O][M] \tag{2.63}$$

where A_f, b_f, and C_f are the rate constants of the reaction; T is the temperature inside a bubble; $[H_2O]$ is the concentration of H_2O molecules inside a bubble; and $[M]$ is the concentration of any molecules (third body) inside a bubble. The subscript f denotes forward reaction. The rate constants of the chemical reactions inside a bubble are listed in Refs. [14, 17, 35, 44, 45].

Rate of the backward reaction of (2.62) can be calculated in a similar manner.

$$r_b = A_b T^{b_b} e^{-C_b/T} [H][OH][M] \tag{2.64}$$

where A_b, b_b, and C_b are the rate constants of the backward reaction, and [H] and [OH] are concentrations of H and OH molecules inside a bubble, respectively. The subscript b denotes backward reaction. The rate constants for the backward reactions are also listed in Refs. [14, 17, 35, 44, 45].

2.9 Single-Bubble Sonochemistry

It is possible to validate a model for bubble dynamics by comparing it with the experimental results on single-bubble sonochemistry [46]. Experimental setup is similar to that of single-bubble sonoluminescence (Fig. 2.9) [22]. A stable single bubble is under controlled acoustic pressure and frequency as well as under controlled liquid temperature. Thus, it is possible to directly compare results of

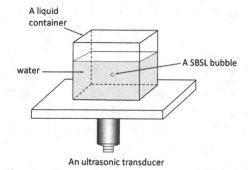

Fig. 2.9 Experimental setup for single-bubble sonochemistry or sonoluminescence (SBSL). Reprinted with permission from Yasui et al. [22]. Copyright (2004), Taylor & Francis

numerical simulation with the experimental data. In this system, there is no complexity due to other bubbles such as bubble–bubble interaction (Sect. 2.18).

In 2002, Didenko and Suslick [46] first reported quantitative results on single-bubble sonochemistry. They measured production rate of OH radicals from a single stable bubble using terephthalate dosimetry (as 8.2×10^5 molecules per acoustic cycle). The ultrasonic frequency and pressure amplitude were 52 kHz and 1.5 atm (=1.52 bar), respectively. The liquid temperature was 3 °C, and the maximum bubble radius was measured as 30.5 µm. The rate of NO_2^- ion production was measured as 9.9×10^6 ions per acoustic cycle, and the number of photons emitted in sonoluminescence was measured as 7.5×10^4.

Next, results of numerical simulation under the experimental condition of Didenko and Suslick [46] are briefly reviewed [17]. To produce the experimentally observed maximum radius of 30.5 µm, the ambient bubble radius was determined as $R_0 = 3.6$ µm (Fig. 2.10a) [17]. The dissolution of OH radicals into surrounding liquid was also numerically simulated by using uptake coefficient Θ defined as follows.

$$\Theta = \frac{N_{in} - N_{out}}{N_{col}} \tag{2.65}$$

where N_{in} is the number of molecules dissolving into the liquid, N_{out} is the number of molecules desorbing from the liquid into the gas, and N_{col} is the number of molecules colliding with the interface between gas and liquid. In the simulation in Ref. [17], the uptake coefficient was assumed to be $\Theta = 0.001$. The rate ($r_{d,OH}$) of dissolution of OH radicals into the liquid water from the interior of a bubble was calculated as follows.

$$r_{d,OH} = \Theta \sqrt{\frac{kT_B}{2\pi m_{OH}}} \frac{n_{OH}}{V} \times 4\pi R^2 \tag{2.66}$$

where k is Boltzmann constant (=1.38×10^{-23} J/K), T_B is the temperature at the bubble wall inside a bubble, m_{OH} is the molecular mass of OH radical (=2.82×10^{-26} kg), n_{OH} is the number of OH radicals inside a bubble, and V is the

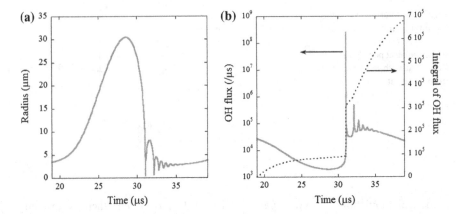

Fig. 2.10 Calculated results for one acoustic cycle when a SBSL bubble in a steady state in water at 3 °C is irradiated by an ultrasonic wave of 52 kHz and 1.52 bar in frequency and pressure amplitude, respectively. The ambient bubble radius is 3.6 μm. **a** The bubble radius. **b** The dissolution rate of OH radicals into the liquid from the interior of the bubble (solid line) and its time integral (dotted line). Reprinted with permission from Yasui et al. [17]. Copyright (2005), AIP Publishing LLC

volume of a bubble ($= \frac{4}{3} \pi R^3$). Equation (2.66) is the frequency of collision of OH radicals on the bubble surface multiplied by surface area of a bubble and uptake coefficient.

The result of numerical simulations on OH flux ($r_{d,OH}$) is shown in Fig. 2.10b as a function of time for one acoustic cycle. The dotted line shows the time integral of the OH flux. The amount of OH radicals dissolved into the surrounding liquid from the interior of a bubble is calculated as 6.6×10^5 (number of molecules). It roughly agrees with the experimental data of 8.2×10^5. This finding means that the model of bubble dynamics including chemical kinetic model is almost validated.

The temperature inside a bubble increases up to 10,900 K at the end of the bubble collapse according to the numerical simulations (Fig. 2.11a). As a result, many chemical species are dissociated at the end of the bubble collapse (Fig. 2.11b). After the end of the bubble collapse (after the time for the minimum radius of a bubble), many chemical species are formed inside a bubble such as H_2, O, H_2O_2, HNO_2, and OH. In the numerical simulations, major non-condensable gas component is assumed as argon based on the argon rectification hypothesis discussed in Sect. 2.4. The amount of N_2 and O_2 inside a bubble is determined by the condition that the amount of N_2 and O_2 diffusing into a bubble by rectified diffusion balances with that dissociated inside a bubble. The chemical species present before the end of the bubble collapse in Fig. 2.11b are generated in the previous violent collapse of a stably pulsating bubble.

The amount of chemical products that dissolve into the liquid water from the interior of a SBSL bubble per acoustic cycle is listed in Table 2.1 according to the numerical simulations [17]. The dominant products are H_2, O, H_2O_2, H, HNO_2,

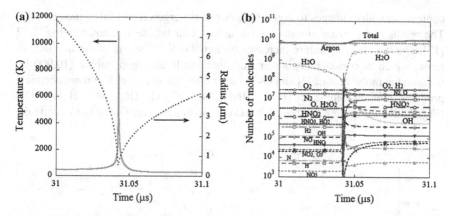

Fig. 2.11 Calculated results for a SBSL bubble in a steady state at around the end of the bubble collapse only for 0.1 μs. **a** The bubble radius and the temperature inside a bubble. **b** The number of molecules inside a bubble. Reprinted with permission from Yasui et al. [17]. Copyright (2005), AIP Publishing LLC

Table 2.1 Amount of chemical products that dissolve into the liquid water from the interior of a SBSL bubble in a steady state per acoustic cycle according to the numerical simulation

Chemical species	Number of molecules per acoustic cycle
H_2	3.1×10^7
O	1.3×10^7
H_2O_2	6.3×10^6
H	4.1×10^6
HNO_2	2.3×10^6
HO_2	1.1×10^6
HNO_3	8.4×10^5
OH	6.6×10^5
NO	2.5×10^5
HNO	9.5×10^4
NO_2	4.4×10^4
O_3	3.4×10^4
N	2.9×10^4
NO_3	3.1×10^3
N_2O	3.1×10^2

Reprinted with permission from Yasui et al. [17]. Copyright (2005), AIP Publishing LLC

HO_2, HNO_3, and OH, and the dominant oxidants are O, H_2O_2, and OH. The main oxidants in sonochemical reactions are discussed in the next section. The amount of HNO_2 of 2.3×10^6 according to the numerical simulation is considerably lower than the experimental value for NO_2^- ions (9.9×10^6) [46]. Further studies are required on this topic.

At the beginning of SBSL experiment, a bubble initially consists of air (N_2, O_2, and Ar) and water vapor. In about one hundred (100) acoustic cycles, the bubble content gradually changes to argon (Ar) due to the argon rectification process [47]. The results of numerical simulations on an initial air bubble are shown in Fig. 2.12 [17]. The bubble temperature increases up to 6500 K at the end of the bubble collapse, which is considerably lower than that inside an argon bubble (10,900 K) because the molar heat of N_2 and O_2 is larger than that of argon. The main chemical products in an air bubble are HNO_2, HNO_3, O, H_2O_2, O_3, HO_2, NO_3, H_2, and OH (Fig. 2.12b, Table 2.2), and the main oxidants in this case are O, H_2O_2, O_3, and OH.

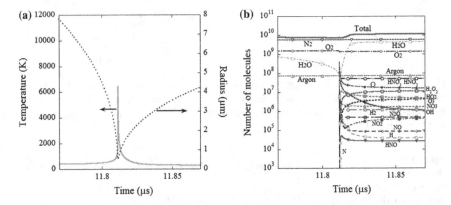

Fig. 2.12 Calculated results for an initial air bubble at around the end of the bubble collapse (only for 0.1 μs). **a** The bubble radius and the temperature inside a bubble. **b** The number of molecules inside a bubble. Reprinted with permission from Yasui et al. [17]. Copyright (2005), AIP Publishing LLC

Table 2.2 Amount of chemical products that dissolve into the liquid water from the interior of an initial air bubble in one acoustic cycle according to the numerical simulation

Chemical species	Number of molecules per acoustic cycle
HNO_2	4.0×10^7
HNO_3	3.7×10^7
O	1.6×10^7
H_2O_2	5.1×10^6
O_3	2.7×10^6
HO_2	2.3×10^6
NO_3	1.1×10^6
H_2	1.0×10^6
OH	9.9×10^5
NO_2	3.9×10^5
N_2O	3.0×10^5
NO	1.3×10^5
H	1.1×10^5
HNO	2.8×10^4
N	2.7×10^3
N_2O_5	6.8×10^2

Reprinted with permission from Yasui et al. [17]. Copyright (2005), AIP Publishing LLC

2.10   Main Oxidants

In order to study the main oxidants created inside an air bubble, numerical simulations of chemical reactions inside a pulsating bubble have been performed for several ultrasonic frequencies and pressure amplitudes [48]. The maximum bubble temperature at the end of the violent bubble collapse is plotted as a function of acoustic pressure amplitude for various ultrasonic frequencies, as shown in Fig. 2.13a [48]. For relatively low ultrasonic frequencies such as 20 and 100 kHz, a peak in the bubble temperature at relatively low acoustic amplitude is observed. The decrease in the bubble temperature as acoustic amplitude increases is due to the increase in vapor fraction inside a bubble at the end of the bubble collapse (Fig. 2.13b) [48]. Vapor fraction increases as the acoustic amplitude increases due to the larger amount of water vapor evaporating into the bubble as the bubble expands more. Because of the larger amount of water vapor at maximum bubble radius, more amount of water vapor is trapped inside a bubble at the end of the bubble collapse due to non-equilibrium condensation during the violent bubble collapse. Larger vapor fraction results in further decrease in bubble temperature because the molar heat of water vapor is larger than that of air and endothermic dissociation of water vapor inside a bubble considerably cools down the bubble. The increase of bubble temperature at low acoustic amplitude is just a result of more expansion of a bubble resulting in more violent collapse. At higher ultrasonic frequencies such as 300 kHz and 1 MHz, bubble temperature continuously increases as the acoustic amplitude increases and reaches a plateau at relatively high acoustic amplitudes. This is due to much lower vapor fraction than that at lower

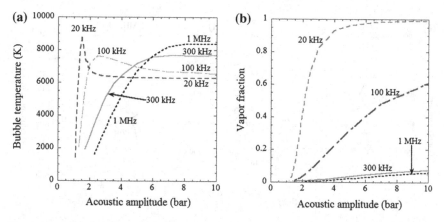

Fig. 2.13 Calculated result as a function of the acoustic amplitude for several ultrasonic frequencies (20, 100, 300 kHz, and 1 MHz) for the first collapse of an isolated spherical air bubble. The ambient bubble radii are 5 μm for 20 kHz, 3.5 μm for 100 and 300 kHz, and 1 μm for 1 MHz. **a** The temperature inside a bubble at the final stage of the bubble collapse. **b** The molar fraction of the water vapor inside a bubble at the end of the bubble collapse. Reprinted with permission from Yasui et al. [48]. Copyright (2007), AIP Publishing LLC

ultrasonic frequencies, which is caused by much smaller expansion of a bubble due
to shorter period of ultrasound.

The quantity of each chemical species becomes nearly constant after about 0.05–
0.1 μs after the end of the bubble collapse according to the numerical simulations
shown in Figs. 2.11b and 2.12b. In Fig. 2.14, the rate of production of each oxidant
inside an air bubble is estimated by the amount of each oxidant inside a bubble at
about 0.05–0.1 μs after the end of the first violent collapse of the bubble. From
Fig. 2.14, the quantity of oxidants at higher bubble temperatures than about 7000 K
is a few orders of magnitude smaller than those at lower bubble temperatures. This
is due to the consumption of oxidants inside an air bubble by oxidizing nitrogen at
higher temperature than about 7000 K. Thus, there is an optimal bubble temperature for oxidant production, which is about 5500 K (Fig. 2.15a) [49]. If nitrogen

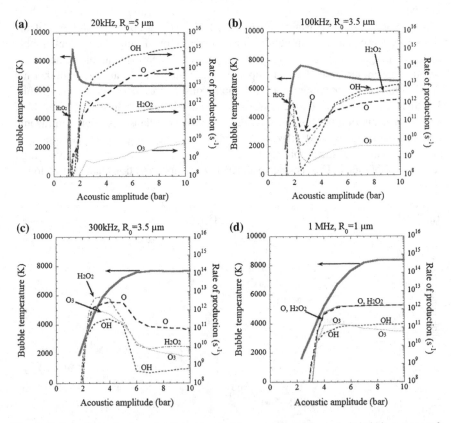

Fig. 2.14 Rate of production of each oxidant inside an isolated spherical air bubble per second
calculated by the first bubble collapse as a function of acoustic amplitude with the temperature
inside a bubble at the end of the bubble collapse (the thick line): **a** 20 kHz and $R_0 = 5$ μm.
b 100 kHz and $R_0 = 3.5$ μm. **c** 300 kHz and $R_0 = 3.5$ μm. **d** 1 MHz and $R_0 = 1$ μm. Reprinted
with permission from Yasui et al. [48]. Copyright (2007), AIP Publishing LLC

Fig. 2.15 Correlation
between the bubble
temperature at the collapse
and the amount of the
oxidants created inside a
bubble per collapse. The
amount of oxidants is in
number of molecules. The
calculated results for various
ambient static pressures and
various acoustic amplitudes
are plotted. The ambient static
pressures are indicated with
the symbols. **a** For an air
bubble with $R_0 = 5$ μm under
ultrasound of 140 kHz. **b** For
an oxygen bubble with
$R_0 = 0.5$ μm under ultrasound
of 1 MHz. Reprinted with
permission from Yasui et al.
[49]. Copyright (2004),
Elsevier

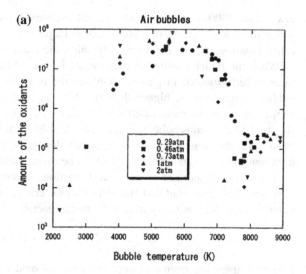

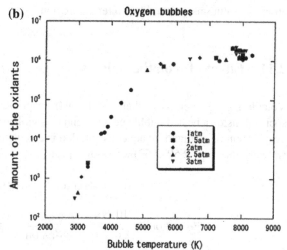

(N_2) is absent, the situation completely changes such as the interior of an oxygen (O_2) bubble because the oxidants are no longer consumed there. For an oxygen bubble, the amount of oxidants created inside a bubble continuously increases as the bubble temperature increases and finally reaches a plateau at higher bubble temperatures than ~ 6000 K (Fig. 2.15b) [49].

When the molar fraction of water vapor inside a bubble at the end of bubble collapse is larger than 0.5, the main oxidant is OH radical because there is a large amount of water vapor as the source of OH radicals inside a bubble (Figs. 2.13b and 2.14a, b) [48]. Furthermore, at these conditions, the bubble temperature is lower than 7000 K, and oxidants are not excessively consumed inside a bubble. When the molar fraction of water vapor is larger than 0.5, a bubble is sometimes called *vaporous*. Thus, in vaporous bubbles, the main oxidant is OH radical.

When the molar fraction of water vapor is smaller than 0.5, a bubble is sometimes called *gaseous*. In gaseous bubbles, the main oxidant is O atom when the bubble temperature is higher than 6500 K (Fig. 2.14) [48]. When the bubble temperature is in the range of 4000–6500 K, the main oxidant is hydrogen peroxide (H_2O_2) in the gaseous bubbles (Fig. 2.14). At 1 MHz, however, H_2O_2 is one of the main oxidants even at temperatures higher than 6500 K because the duration of high temperature is too short for H_2O_2 to be dissociated inside a bubble.

The role of O atom in sonochemical reactions in solutions is still under debate [33] (see Sect. 3.9). Hart and Henglein [50] experimentally suggested that O atom contributes to KI dosimetry in sonochemical reactions as follows.

$$O + 2I^- + 2H + \rightarrow I_2 + H_2O \qquad (2.67)$$

However, there has been no direct experimental evidence on the production of O atoms in solutions (at liquid or interface regions).

2.11 Effect of Volatile Solutes

Volatile solutes such as methanol (CH_3OH) evaporate into a bubble and are dissociated inside a heated bubble near the end of a violent bubble collapse. The rate of evaporation of methanol in aqueous solution is calculated in a similar way to that of water vapor (Eqs. 2.50–2.52) as follows [51].

$$\dot{m}_{CH_3OH} = \dot{m}_{eva,CH_3OH} - \dot{m}_{con,CH_3OH} \qquad (2.68)$$

$$\dot{m}_{eva,CH_3OH} = \frac{10^3 N_A}{M_{CH_3OH}} \frac{\alpha_{M,CH_3OH}}{\sqrt{2\pi R_{CH_3OH}}} \frac{p^*_{CH_3OH}}{\sqrt{T_{L,i}}} \acute{a} S_{CH_3OH} \qquad (2.69)$$

$$\dot{m}_{con,CH_3OH} = \frac{10^3 N_A}{M_{CH_3OH}} \frac{\alpha_{M,CH_3OH}}{\sqrt{2\pi R_{CH_3OH}}} \frac{\Gamma_{CH_3OH} p_{CH_3OH}}{\sqrt{T_B}} \qquad (2.70)$$

where M_{CH_3OH} is molecular weight of methanol (=32 g/mol), α_{M,CH_3OH} is accommodation coefficient for the evaporation or condensation for methanol, R_{CH_3OH} is the gas constant for methanol (=260 J/kg K), $p^*_{CH_3OH}$ is the saturated vapor pressure of methanol at liquid temperature ($T_{L,i}$) at the bubble wall, $\acute{a}$ is the area occupied by a methanol molecule at the gas/liquid interface (=2 × 10^{-19} m²/molecule), S_{CH_3OH} is the instantaneous surface concentration of methanol at the bubble wall, p_{CH_3OH} is the partial pressure of methanol inside a bubble, and the correction factor (Γ_{CH_3OH}) is calculated in a similar way to that of water vapor (Eqs. 2.53 and 2.54) as follows.

$$\Gamma_{CH_3OH} = e^{-\Omega_{CH_3OH}^2} - \Omega_{CH_3OH}\sqrt{\pi}\left[1 - \frac{2}{\sqrt{\pi}}\int_0^{\Omega_{CH_3OH}} e^{-x^2}dx\right] \qquad (2.71)$$

where

$$\Omega_{CH_3OH} = \frac{\dot{m}_{CH_3OH}}{p_{CH_3OH}}\left(\frac{R_{CH_3OH}T}{2}\right)^{1/2} \qquad (2.72)$$

Saturated vapor pressure of methanol is calculated as a function of the liquid temperature at the bubble wall [51]. The accommodation coefficient for methanol is assumed to be the same as that for water vapor [51].

The surface concentration of methanol at the bubble wall is calculated as follows.

$$S_{CH_3OH} = \frac{N_{S,CH_3OH}}{4\pi R^2} \qquad (2.73)$$

where N_{S,CH_3OH} is the instantaneous number of methanol molecules sitting at the bubble wall. The change in number of methanol molecules at the bubble wall in time Δt is given as follows.

$$\Delta N_{S,CH_3OH} = -4\pi R^2 \dot{m}_{CH_3OH}\Delta t + \Delta N_{diff,CH_3OH} \qquad (2.74)$$

where the first term is the change by evaporation or condensation at the bubble wall, and the second term is the diffusion of methanol molecules in liquid. The second term is calculated in a similar way to that for the diffusion of a non-condensable gas (Eq. 2.60) [51].

$$\Delta N_{diff,CH_3OH} = 4\pi R^2 D_{CH_3OH}\frac{A_R}{B_R}\frac{c_{\infty,CH_3OH} - c_{i,CH_3OH}}{(R_0/R)^2 R_0}\Delta t \qquad (2.75)$$

where D_{CH_3OH} is the diffusion coefficient of methanol in the liquid water, c_{∞,CH_3OH} is the ambient concentration of methanol, and c_{i,CH_3OH} is the instantaneous concentration of methanol near the bubble wall.

The partial coverage of the bubble surface by methanol molecules partially inhibits evaporation and condensation of water at the bubble wall. The effective area for evaporation and condensation of water is reduced from $4\pi R^2$ to $4\pi R^2(1 - \acute{a}S_{CH_3OH})$.

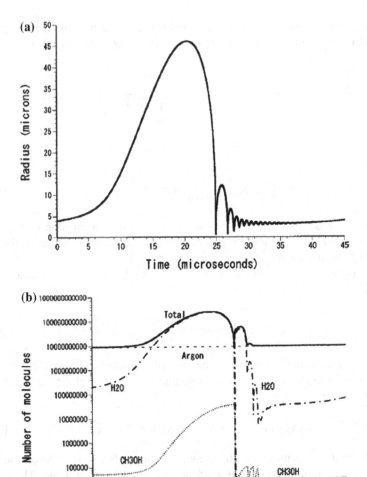

Fig. 2.16 Calculated results of an argon bubble in aqueous methanol solution [$S_{eq} = 0.01 \times 10^{14}$ (molecules/cm^2), where S_{eq} is the equilibrium surface concentration of methanol] as a function of time for one acoustic cycle (45 μs) when the frequency and amplitude of ultrasound are 22 kHz and 1.32 bar, respectively, and the ambient bubble radius is 4 μm. **a** Bubble radius. **b** Number of molecules inside a bubble with logarithmic scale. Reprinted with permission from Yasui [51]. Copyright (2002), AIP Publishing LLC

The change in number of methanol molecules inside a bubble (n_{CH_3OH}) is calculated as follows.

$$n_{CH_3OH}(t + \Delta t) = n_{CH_3OH}(t) + 4\pi R^2 \dot{m}_{CH_3OH}\Delta t - \frac{4}{3}\pi R^3 r_{CH_3OH + M \to CH_3 + OH + M}\Delta t$$

$$(2.76)$$

where $r_{CH_3OH + M \to CH_3 + OH + M}$ is the rate of the following chemical reaction.

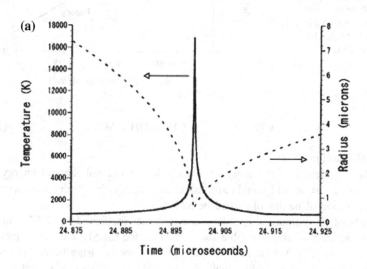

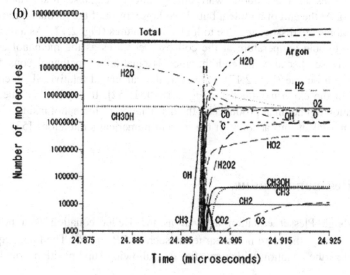

Fig. 2.17 Calculated results at around the minimum bubble radius in aqueous methanol solution. The condition is the same as that in Fig. 2.16. **a** Bubble radius and temperature inside a bubble. **b** Number of molecules inside a bubble with logarithmic scale. Reprinted with permission from Yasui [51]. Copyright (2002), AIP Publishing LLC

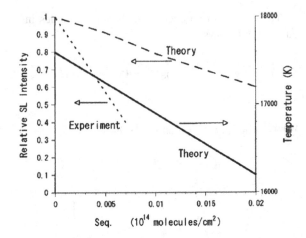

Fig. 2.18 Energy of the emitted light per bubble collapse and the bubble temperature at the collapse as a function of equilibrium surface concentration of methanol (S_{eq}). The experimental data of the relative sonoluminescence intensity (Ref. [52]) are also shown. Reprinted with permission from Yasui [51]. Copyright (2002), AIP Publishing LLC

$$CH_3OH + M \rightarrow CH_3 + OH + M \tag{2.77}$$

where M is a third body.

In the calculation of the temporal change in the internal thermal energy of a bubble (Eq. 2.49), an additional term accounting energy change due to evaporation or condensation of methanol should be added.

Results of the numerical simulations are shown in Figs. 2.16, 2.17, and 2.18 under a condition of single-bubble sonoluminescence (SBSL) and argon rectification (Sect. 2.4) [51]. Methanol evaporates into a bubble during bubble expansion and condenses at the bubble wall during bubble collapse like water vapor (Fig. 2.16). At the end of a violent bubble collapse, most of the methanol molecules inside a bubble are dissociated due to high temperatures (Fig. 2.17). As a result, the maximum bubble temperature at the collapse decreases as the methanol concentration increases because the endothermic dissociation of methanol considerably cools down a bubble (Fig. 2.17). However, the calculated relative SL intensity is higher than the experimental data (Fig. 2.18) [51, 52]. It is probably due to the accumulation of chemical products by the dissociation of methanol inside a bubble, which is not taken into account in the present numerical simulations [53].

2.12 Resonance Radius

The Rayleigh–Plesset equation as well as the Keller equation is a nonlinear equation because there are nonlinear terms such as $R\ddot{R}$ and $\dot{R}^2$. The linear equation is defined as the equation which satisfies the following superposition principle.

$$f(\alpha x_1 + \beta x_2) = \alpha f(x_1) + \beta f(x_2) \tag{2.78}$$

where f is a function of x, x_1 and x_2 are arbitrary values of x, and α and β are arbitrary constants. If a function f does not satisfy Eq. (2.78), it is a nonlinear function. For example, a function f defined by $f(R) = R + \ddot{R}$ is a linear function because $f(\alpha R_1 + \beta R_2) = \alpha f(R_1) + \beta f(R_2)$. On the other hand, a function f defined by $f(R) = R\ddot{R}$ is a nonlinear function because

$$f(\alpha R_1 + \beta R_2) = (\alpha R_1 + \beta R_2) \cdot (\alpha \ddot{R}_1 + \beta \ddot{R}_2) \neq \alpha f(R_1) + \beta f(R_2) \tag{2.79}$$

Generally speaking, a linear function of x consists only of a first-order term (ax, where a is a constant) and a constant. Other functions are nonlinear such as a function containing higher order terms (bx^2, cx^3, etc.).

Any radius–time curve ($R(t)$) can be expressed by the non-dimensional function $x(t)$ defined as follows.

$$R(t) = R_0(1 + x(t)) \tag{2.80}$$

$x(t)$ means degree of deviation of $R(t)$ from the ambient value R_0. Now let us consider very weak pulsation of a bubble so that $|x(t)| \ll 1$. In this case, $1 \gg |x(t)| \gg |x(t)|^2$ holds. Thus, in the Rayleigh–Plesset equation, only the first-order terms need to be considered. Inserting Eq. (2.80) into the Rayleigh–Plesset equation (Eq. 2.12) gives the following equation (by neglecting the higher order terms).

$$\ddot{x} + 2\gamma_d \dot{x} + \omega_0^2 x = f_0 \sin \omega t \tag{2.81}$$

where

$$\gamma_d = \frac{2\mu}{\rho_0 R_0^2} \tag{2.82}$$

$$\omega_0 = \frac{1}{R_0} \sqrt{\frac{\left[3\gamma p_0 + (3\gamma - 1)\frac{2\sigma}{R_0}\right]}{\rho_0}} \tag{2.83}$$

$$f_0 = \frac{A}{\rho_0 R_0^2} \tag{2.84}$$

In the derivation, the bubble pulsation is assumed to be adiabatic as follows.

$$p_g(t) = \left(p_0 + \frac{2\sigma}{R_0}\right)\left(\frac{R}{R_0}\right)^{-3\gamma} \tag{2.85}$$

where γ is ratio of specific heats ($\gamma = 1.4$ for air, 1.67 for argon). Furthermore, the vapor pressure (p_v) and the Laplace pressure $\left(\frac{2\sigma}{R_0}\right)$ are neglected compared to the acoustic pressure amplitude (A). Equation (2.81) is an equation for a forced oscillator with a natural frequency ω_0 and with damping constant γ_d.

Solution of Eq. (2.81) is given as follows [54]:

$$x = ae^{-\gamma_d t}\cos(\omega_\gamma + \alpha) + \frac{f_0}{\sqrt{(\omega_0^2 - \omega^2)^2 + (2\gamma_d\omega)^2}}\sin(\omega t + \varphi) \qquad (2.86)$$

where a is a constant. $\omega_\gamma = \sqrt{\omega_0^2 - \gamma_d^2}$, where $\omega_0 > \gamma_d$ is used. α is a constant and φ is given as follows.

$$\varphi = \tan^{-1}\frac{-2\gamma_d\omega}{\omega_0^2 - \omega^2} = \tan^{-1}\left[\frac{-2F\Gamma_\gamma}{(1 - F^2)}\right] \qquad (2.87)$$

where F and Γ_γ are defined as follows.

$$F = \frac{\omega}{\omega_0} \qquad (2.88)$$

$$\Gamma_\gamma = \frac{\gamma_d}{\omega_0} \qquad (2.89)$$

The first term on right-hand side of Eq. (2.86) becomes negligible after a sufficient time as $e^{-\gamma_d t} \to 0$ with $t \to \infty$. Then, the amplitude of oscillation (A_1) is given as follows.

$$A_1 = \frac{A_0}{\sqrt{(1 - F^2)^2 + 4\Gamma_\gamma^2 F^2}} \qquad (2.90)$$

where $A_0 = \frac{f_0}{\omega_0^2}$.

The results of the numerical calculations of Eqs. (2.87) and (2.90) are shown in Fig. 2.19 for an air bubble in water at 20 °C. The amplitude of oscillation has a sharp peak at $\omega = \omega_0$ for both the ambient radii of 1 and 1000 μm. Thus, the frequency given by Eq. (2.83) is called (linear) resonance frequency ($= \frac{\omega_0}{2\pi}$) which is shown in Fig. 2.20. The resonance frequency decreases as the ambient bubble radius (R_0) increases from 4750 kHz at $R_0 = 1$ μm to 3.29 kHz at $R_0 = 1000$ μm. As the effect of viscosity (damping) becomes weaker as the ambient bubble radius increases, the peak in amplitude of oscillation becomes higher (Fig. 2.19a). With regard to the phase difference between the driving ultrasound and the bubble oscillation (Eq. 2.87 in Fig. 2.19b), ultrasound and the bubble oscillation are in phase (no phase difference) when the driving frequency is much lower than the resonance frequency of a bubble ($F \sim 0$). On the other hand, ultrasound and bubble

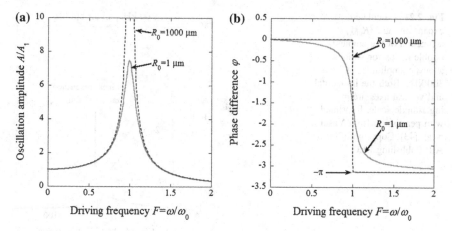

Fig. 2.19 Oscillation amplitude (Eq. 2.90) (**a**) and phase difference (Eq. 2.87) (**b**) as a function of the driving frequency in a forced oscillator (bubble pulsation)

Fig. 2.20 Linear resonance frequency of a bubble as a function of the ambient radius

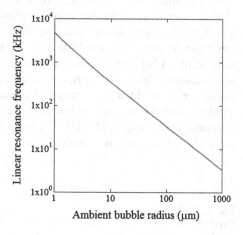

oscillation are in anti-phase (a phase difference of $-\pi$) when the driving frequency is much higher than the resonance frequency ($F \gg 1$). For larger ambient bubble radius, this behavior is more significant due to negligible effect of damping (Fig. 2.19b).

However, the bubble pulsation in acoustic cavitation is strongly nonlinear. The linear approximation is valid only for very low acoustic amplitudes. As shown in Fig. 2.21, the peak in the expansion ratio (R_{max}/R_0, where R_{max} is the maximum radius of a pulsating bubble) is at about $R_0 = 6$ μm for an acoustic amplitude of 0.5 bar (= 5×10^4 Pa = 0.493 atm) at 300 kHz, which is considerably smaller than the linear resonance radius of 11.4 μm (Eq. 2.83) [43]. For an acoustic amplitude of 3 bar, the peak is at about 0.4 μm, which is more than one order of magnitude smaller than the linear resonance radius. For this acoustic amplitude, a

Fig. 2.21 Calculated expansion ratio (R_{max}/R_0) as a function of the ambient bubble radius for various acoustic amplitudes at 300 kHz. Both the horizontal and vertical axes are in logarithmic scale. Reprinted with permission from Yasui et al. [43]. Copyright (2008), AIP Publishing LLC

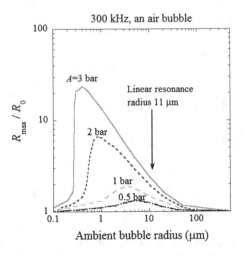

bubble significantly expands $(R_{max}/R_0 > 3)$ in the range of 0.27 μm < R_0 < 7 μm [43]. In other words, a bubble actively pulsates for a wide range of ambient radius which does *not* include the linear resonance radius of 11.4 μm. This means that at this acoustic amplitude, a bubble of linear resonance size is *inactive* due to the nonlinearity of bubble pulsation. The ambient radius for the peak in expansion ratio is sometimes called *nonlinear* resonance radius. The peak is not sharp, and a bubble strongly pulsates in a wide range of ambient radius. The upper bound of the ambient radius for an active bubble is in the same order of magnitude as the linear resonance radius (Fig. 2.22) [43]. The lower bound nearly coincides with the *Blake threshold* radius given as follows [2, 39].

Fig. 2.22 Calculated thresholds for shape instability, sonoluminescence (SL), and oxidant production as well as the Blake threshold (Eq. 2.91) in $R_0 - A$ plane. Reprinted with permission from Yasui et al. [43]. Copyright (2008), AIP Publishing LLC

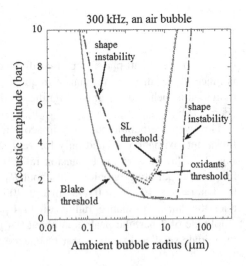

$$p_{\text{Blake}} = p_0 + \frac{8\sigma}{9}\sqrt{\frac{3\sigma}{2R_{0,\text{Blake}}^3\left[p_0 + (2\sigma/R_{0,\text{Blake}})\right]}} \qquad (2.91)$$

where p_{Blake} is the pressure amplitude of ultrasound (the Blake threshold pressure for transient cavitation), and $R_{0,\text{Blake}}$ is the ambient bubble radius (the Blake threshold radius). The Blake threshold pressure is a threshold above which a bubble significantly expands (and subsequently collapses). The derivation of Eq. (2.91) is given in Ref. [2].

2.13 Shock Wave Emission

The Mach number (M) is defined as the fluid velocity divided by the sound velocity. A shock wave is a propagation of a discontinuity in pressure in a medium [10, 55]. For a shock wave propagating in a homogeneous medium under steady-state conditions, upstream velocity of fluid is normally higher than sound velocity ($M > 1$) according to the *Rankine–Hugoniot* relations [10]. It is also known that a shock wave is emitted when the velocity of an airplane exceeds sound velocity. Generally speaking however, shock wave can be emitted even when the Mach number is less than 1. An example is shock wave emission from a bubble at its collapse. Figure 2.23 shows a photograph of a spherical shock wave emitted from a bubble at its collapse [56]. The circle in Fig. 2.23 is the spherical shock wave propagating outward from the center of a circle where a bubble is present although it is invisible in the photograph.

The reason for the shock wave emission from a bubble can be understood from a result of numerical simulations (Figs. 2.24 and 2.25) [57]. As shown in Fig. 2.24a, b, the shock formation occurs during rebounding of a bubble after a violent collapse.

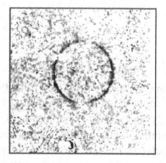

Fig. 2.23 An image of a spherical shock wave emitted by a sonoluminescing bubble. At 480 ns after the bubble collapse. The side length of the image is 3.5 mm. Reprinted with permission from Holzfuss et al. [56]. Copyright (1998), American Physical Society

Fig. 2.24 Variation of
pressure with distance from
the bubble wall at various
instances in time. **a** During
the collapse of a bubble.
b During the rebound of a
bubble. Reprinted with
permission from Hickling and
Plesset [57]. Copyright
(1964), AIP Publishing LLC

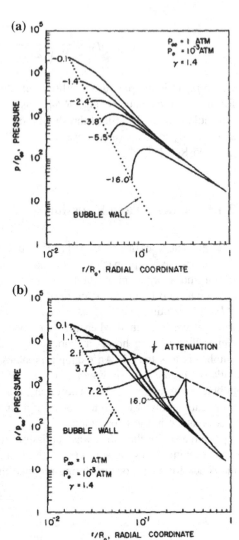

Just after a violent collapse, a bubble starts expanding. However, the liquid slightly
apart from a bubble still flows toward a bubble (the velocity is still negative) (e.g., at
time 0.1 after the end of the collapse in Fig. 2.25, the fluid velocity (Mach number) is
negative at a region slightly away from a bubble wall). Furthermore, the sound
velocity decreases as the distance from a bubble increases because the liquid pressure
decreases (the sound velocity decreases as the liquid pressure decreases). The
pressure wave continuously emitted from an expanding bubble propagates with the
speed equivalent to the sum of the sound velocity and the fluid (liquid) velocity. Due
to the above two factors, the pressure waves radiated from an expanding bubble

Fig. 2.25 Variation of the
Mach number with distance
from the bubble wall at
different instances in time
during the collapse and
rebound of a bubble.
Reprinted with permission
from Hickling and Plesset
[57]. Copyright (1964), AIP
Publishing LLC

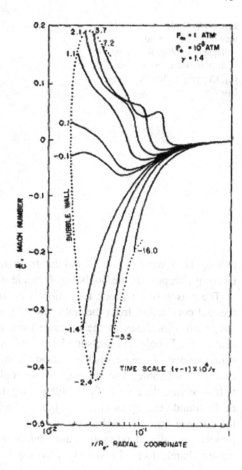

overtake the previously radiated pressure waves. In this way, a shock wave is formed
as seen as a pressure peak propagating outward as shown in Fig. 2.24b. It should be
noted that the absolute value of the Mach number of the bubble wall is much less than
1 during the formation of the shock wave (Fig. 2.25) [54].

2.14 Shock Formation Inside a Bubble

According to some numerical simulations of fundamental equations of fluid
dynamics *inside* a collapsing bubble, a spherical shock wave is formed, which
propagates inwardly and finally focuses at the bubble center [58–60]. When a
spherical shock wave focuses at the bubble center, temperature increases to about
10^6 K (1,000,000 K) or more. However, other numerical simulations taking into
account the effect of thermal conduction inside a collapsing bubble have revealed

Fig. 2.26 Difficulty in shock wave formation inside a collapsing bubble. Reprinted with permission from Yasui [2]. Copyright (2015), Elsevier

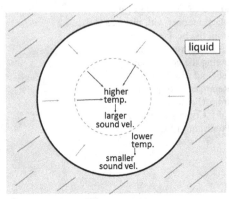

A collapsing bubble

that no shock wave is formed and that temperature and pressure are almost spatially uniform except near the bubble wall inside a collapsing bubble [61–65].

The reason of no shock formation is as follows (Fig. 2.26) [2, 66]. Due to the thermal conduction from the hotter bubble interior to the colder bubble wall, the temperature increases as the distance from the bubble wall increases toward the center of a bubble. As the sound velocity increases as temperature increases, the sound velocity increases as the distance from the bubble wall increases toward the center of a bubble. The pressure waves radiated from the bubble wall inside a bubble toward the center of a bubble propagate with the speed equivalent to the sum of the sound velocity and the fluid (gas) velocity. As the sound velocity increases as the distance from the bubble wall increases, the pressure waves barely overtake the previously radiated pressure waves inside a collapsing bubble. Thus, the shock wave is barely formed inside a collapsing bubble.

2.15 Jet Penetration Inside a Bubble

When the pressure field around a bubble is strongly asymmetric, a liquid jet penetrates into a collapsing bubble. It is most often observed for the bubble collapse near a solid surface (Figs. 2.27 and 2.28) [67, 68]. Just before the bubble collapse near a solid surface, the liquid pressure on the bubble surface near a solid boundary becomes much lower than that on the other side of the bubble surface (Fig. 2.28) [68]. As a result, a jet penetrates into a bubble and finally hits the solid surface. Then, the liquid flow spreads on the solid surface and cleans the surface (contaminated with small particles) (Fig. 2.29d) [67]. During the bubble expansion (Fig. 2.29a) and bubble collapse (Fig. 2.29b), fluorescent particles of 8 μm in diameter initially distributed on the solid surface moved associated with the liquid flow due to bubble pulsation. However, the permanent displacement of particles

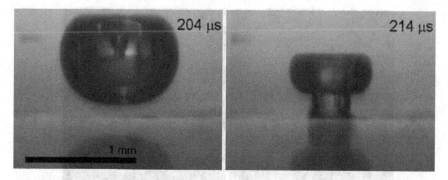

Fig. 2.27 Stroboscopic pictures visualizing liquid jet within a bubble (left) and impacting on the solid boundary (right). The bubble was generated by focusing an intense laser pulse of 1064 nm in wavelength and 6 ns in duration. Reprinted with permission from Ohl et al. [67]. Copyright (2006), AIP Publishing LLC

Fig. 2.28 Results of numerical simulations for the collapse of an initially spherical bubble near a plane solid wall when the bubble boundary was in contact with the solid wall. Reprinted with permission from Plesset and Chapman [68]. Copyright (1971), Cambridge University Press

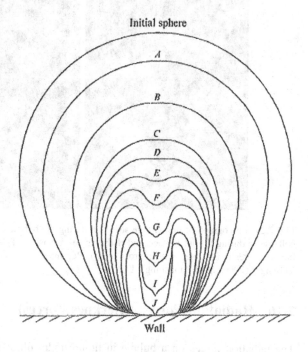

was observed only during the jet impact (from Fig. 2.29c, d) [67]. A jet impact also causes a damage of the solid surface (Fig. 2.30) [69].

Jet penetration also occurs when two neighboring bubbles simultaneously collapse (Fig. 2.31) [70]. Another situation of jet penetration into a bubble is the bubble collapse in a traveling ultrasonic wave [71].

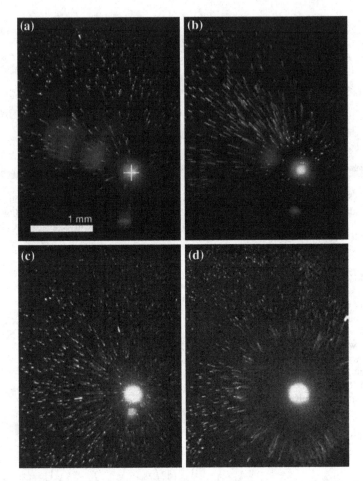

Fig. 2.29 Photograph of the particle streaks **a** during the bubble expansion, **b** during the bubble collapse, **c** during the jet impact, and **d** during re-expansion of the bubble. The position of the bubble center is indicated with a cross in (**a**). Reprinted with permission from Ohl et al. [67]. Copyright (2006), AIP Publishing LLC

2.16   Radiation Forces (Bjerknes Forces)

The radiation forces on a bubble in liquid under ultrasound originate in pressure inhomogeneity around a bubble. If the pressure inhomogeneity originates in an external acoustic field (driving ultrasound), the radiation force is called *primary Bjerknes force*. If it originates in an acoustic wave radiated by a neighboring bubble, it is called *secondary Bjerknes force*. Both primary and secondary Bjerknes forces are principally expressed by the same equation as follows [72].

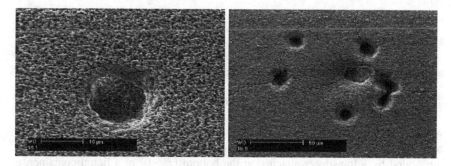

Fig. 2.30 SEM images showing evidence of microjet impacts on the surface of the cake layer made of sulfate polystyrene latex particles of 0.53 μm in average diameter. Left: ultrasound operating at 1062 kHz for 5 s and 0.21 W/cm². Right: ultrasound operating at 620 kHz for 5 s and 0.12 W/cm². Reprinted with permission from Lamminen et al. [69]. Copyright (2004), Elsevier

Fig. 2.31 Comparison between experimental and simulation of the cavitation of two bubbles initially set at a distance of 400 μm subjected to a minimum pressure of −1.4 MPa. Each time step is 4 μs in the direction from the top to the bottom (the total time is 20 μs). Reprinted with permission from Bremond et al. [70]. Copyright (2006), AIP Publishing LLC

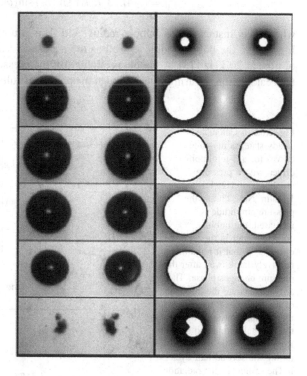

$$\vec{F}_B = \langle \vec{F}_p \rangle = -\langle V \vec{\nabla} p \rangle \qquad (2.92)$$

where $\vec{F}_B$ is the primary or secondary Bjerknes force, $\vec{F}_p$ is the instantaneous radiation force which dramatically changes in one acoustic cycle including direction of the force, $\langle \ \rangle$ means the time-averaged value, V is the instantaneous bubble volume, $\vec{\nabla} = \left(\frac{\partial}{\partial x}, \frac{\partial}{\partial y}, \frac{\partial}{\partial z}\right)$, and $p = p(x, y, z, t)$ is the instantaneous pressure field around a bubble.

Firstly, the primary Bjerknes force in a standing acoustic field is discussed. As an example, the instantaneous pressure field is given as follows.

$$p(z, t) = -A \cos(kz) \sin(\omega t) \tag{2.93}$$

where A is the pressure amplitude of ultrasound, and the origin ($z = 0$, $t = 0$) has been appropriately shifted compared to Eq. (1.28). When a liquid is irradiated with ultrasound by a transducer attached at the bottom ($z = 0$) of a liquid container, a standing wave field is formed as given by Eq. (2.93) if the liquid height is at $z = (2n + 1)\pi/2k$, where z-axis is in the vertical direction and n is a natural number. Then, instantaneous radiation force is given as follows.

$$\vec{F}_p = (-4\pi/3)R^3 kA \sin(kz) \sin(\omega t)\vec{e}_z \tag{2.94}$$

where R is instantaneous bubble radius, and $\vec{e}_z$ is a unit vector in z direction. In Fig. 2.32a, the bubble radius as well as the acoustic pressure is shown as function of time at 20 kHz [73]. A bubble is slightly off the pressure antinode (by 1 mm) in the calculation. During the rarefaction phase of ultrasound, over the initial

Fig. 2.32 a Calculated steady-state radius–time curves for a 5-μm bubble driven with a pressure amplitude of 1.3, 1.5, and 1.7 atm. Note that as the pressure amplitude is increased, the bubble collapses later in the acoustic cycle, such that it remains relatively large even after the pressure changes phase. The arrows illustrate the direction of the Bjerknes force, toward the pressure antinode during the first half cycle and away from the pressure antinode during the second half cycle. b The instantaneous radiation force for bubbles driven as shown in (a). Reprinted with permission from Matula et al. [73]. Copyright (1997), AIP Publishing LLC

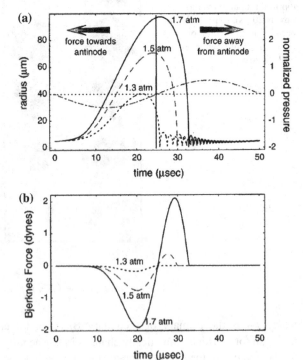

half-wave period (0–25 μs), a bubble expands. At this stage, the instantaneous acoustic pressure is lowest at pressure antinode, and the instantaneous radiation force is directed toward the pressure antinode. [The instantaneous radiation force is directed toward lower pressure region as indicated in Eq. (2.92).] In the compression phase of ultrasound during the latter half-wave period (25–50 μs), a bubble collapses and undergoes bouncing motion. At this stage, the instantaneous acoustic pressure is highest at pressure antinode, and the instantaneous radiation force is directed away from pressure antinode. As the instantaneous radiation force is proportional to bubble volume (Eq. 2.92), the force is stronger during bubble expansion compared to that during bubble collapse (Fig. 2.32b). As a result, time-averaged radiation pressure (primary Bjerkens force) is directed toward the pressure antinode. However, when the acoustic pressure amplitude is higher than about 1.8 bar at 20 kHz, bubble expansion still continues at the beginning of compression phase of ultrasound. This gives rise to a repulsive force from the pressure antinode in the compression phase that is stronger than the attractive one generated during the rarefaction phase. Consequently, a bubble is repelled from the pressure antinode above about 1.8 bar at 20 kHz [73]. This behavior is seen in the experimental observation of the bubble structure in Fig. 1.11.

Next, the primary Bjerknes force in a traveling acoustic wave is discussed. In this case, instantaneous pressure field is given as follows as in Eq. (1.2).

$$p(z,t) = -A(z)\sin(\omega t - kz) \tag{2.95}$$

where the acoustic pressure amplitude A is a function of position z as in the case of an acoustic field under an ultrasonic horn (probe) (Eq. 1.13), and then, the instantaneous radiation force is given as follows.

$$\vec{F}_{\mathrm{p}} = (4\pi/3)R^3 \sin(\omega t - kz)\vec{\nabla}A - (4\pi/3)AkR^3\cos(\omega t - kz)\vec{e}_z \tag{2.96}$$

In many cases, time-averaged $\vec{F}_{\mathrm{p}}$ is directed away from a horn tip [74].

Finally, the secondary Bjerknes force is discussed [75].

$$\vec{F}_{1\rightarrow 2} = -V_2\vec{\nabla}p_1 \tag{2.97}$$

where $\vec{F}_{1\rightarrow 2}$ is the force acting on bubble 2 from bubble 1, V_2 is the volume of bubble 2, and p_1 is the acoustic pressure radiated from bubble 1. The pulsating bubble radiates acoustic wave into the surrounding liquid. Let us consider the Euler equation (equation of motion) in fluid dynamics already discussed in Eq. (2.16).

$$\frac{\partial \vec{u}}{\partial t} + (\vec{u}\cdot\nabla)\vec{u} = -\frac{1}{\rho}\nabla p \tag{2.98}$$

The fluid (liquid) velocity ($\vec{u}$) around a pulsating bubble is given by Eq. (2.99) according to the condition of incompressibility of liquid [described below Eq. (2.2)].

$$\vec{u} = \frac{R^2 \dot{R}}{r^2} \vec{e}_r \tag{2.99}$$

where $\vec{e}_r$ is a radial unit vector with its origin at the center of a bubble. Then, the second term on the left-hand side of Eq. (2.98) is proportional to r^{-5} and negligible compared to the first term. Inserting Eq. (2.99) into Eq. (2.98) yields Eq. (2.100).

$$\frac{\partial p}{\partial r} = -\frac{\rho}{r^2} \frac{d}{dt} \left(R^2 \dot{R} \right) \tag{2.100}$$

where p is the acoustic pressure radiated from a bubble. Integrating Eq. (2.100) with r yields Eq. (2.101).

$$p = \frac{\rho}{r} \frac{d}{dt} \left(R^2 \dot{R} \right) = \frac{\rho}{4\pi r} \frac{d^2 V}{dt^2} \tag{2.101}$$

where $V = \frac{4}{3}\pi R^3$ is the instantaneous volume of a bubble. Inserting Eq. (2.101) into Eq. (2.97) yields Eq. (2.102).

$$\vec{F}_{1 \to 2} = \frac{\rho}{4\pi d^2} \left\langle \ddot{V}_1 V_2 \right\rangle \vec{e}_{1 \to 2} \tag{2.102}$$

where d is the distance between the bubble centers of bubbles 1 and 2, V_1 and V_2 are the volumes of bubbles 1 and 2, respectively, $\ddot{V}_1 = \frac{d^2 V_1}{dt^2}$, and $\vec{e}_{1 \to 2}$ is a unit vector directed from bubble 1 to bubble 2. When the coefficient of $\vec{e}_{1 \to 2}$ in Eq. (2.102) is negative, the secondary Bjerknes force is attractive.

In Fig. 2.33, the coefficient for the secondary Bjerknes force (coefficient of $\vec{e}_{1 \to 2}$) is shown for various combinations of ambient radii of bubbles 1 and 2 at 20 kHz

Fig. 2.33 Coefficient for the secondary Bjerknes force (coefficient of $\vec{e}_{1 \to 2}$). The horizontal and vertical axes are ambient radii of bubble 1 and 2, respectively. The black (white) region shows attractive (repulsive) secondary Bjerknes force. Reprinted with permission from Mettin et al. [75]. Copyright (1997), American Physical Society

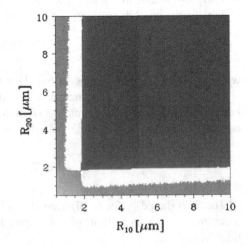

and 1.32 bar according to numerical calculations [75]. The black (white) region shows that the secondary Bjerkens force is attractive (repulsive). For most combinations of ambient radii, the secondary Bjerkens force is attractive. However, when one of the bubbles has a relatively small ambient radius (1–2 μm), the secondary Bjerknes force is repulsive if the ambient radius of the other bubble is larger.

2.17 Effect of Salts and Surfactants

Bubble–bubble coalescence is strongly retarded by the presence of surfactants, salts, alcohols, and sugars (e.g., glucose) above certain concentrations [76–80]. In other words, coalescence of bubbles is largely inhibited if solute concentration is above a critical one which is often called *transition concentration*. Different mechanisms have been proposed to explain the inhibition of coalescence of bubbles in the presence of various solutes [76]. However, there is no consensus on a mechanism which can explain the inhibition of coalescence of bubbles for all such solutes. However, one of many promising mechanisms is described as follows. Let us consider a liquid film between two coalescing bubbles (Fig. 2.34). For coalescence to be completed, a liquid film should be ruptured. In the presence of solutes, thinning of a liquid film which is necessary for its rupture could be inhibited by higher surface tension at center of a film compared to that at the edge of a film. The difference in surface tension originates in gradient of concentration of solutes along the film. The difference in surface tension may be expressed as follows [77, 81–83].

$$\Delta\sigma = -\frac{1}{vh}\left(\frac{2c_{\text{solute}}}{R_g T_L}\right)\left(\frac{\partial\sigma}{\partial c_{\text{solute}}}\right)^2 \tag{2.103}$$

where $\Delta\sigma$ is the surface tension at the edge of a film minus that at the film center, v is the number of ions produced upon dissociation when the solute is electrolyte, h is the thickness of a liquid film (Fig. 2.34), c_{solute} is the solute concentration, R_g is the universal gas constant, and T_L is the liquid temperature. Surface tension at the film center is higher than that at the edge of a film because Eq. (2.103) is always

Fig. 2.34 Two coalescing bubbles

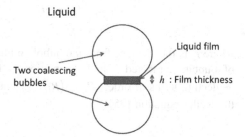

Liquid

Liquid film

Two coalescing bubbles

h : Film thickness

negative. It works against thinning of a liquid film. The magnitude of the difference in surface tension is proportional to $\left(\frac{\partial \sigma}{\partial c_{\text{solute}}}\right)^{2}$. It has been reported that the experimental transition concentration is well correlated with $\left(\frac{\partial \sigma}{\partial c_{\text{solute}}}\right)^{2}$ (or equivalently with $\left(\frac{\partial \sigma}{\partial c_{\text{solute}}}\right)^{-2}$) [80]. Further studies are required on the mechanism of inhibition of bubble coalescence by solutes above transition concentration.

In acoustic cavitation, bubble–bubble coalescence frequently occurs by attractive secondary Bjerknes forces [38]. As a result, larger bubbles are frequently formed, and average bubble size gradually increases although some of the bubbles are fragmented into smaller "daughter" bubbles [84, 85]. In the presence of solutes above a certain concentration, bubble–bubble coalescence is largely inhibited even in acoustic cavitation. As a result, number of tiny active bubbles increases as larger inactive bubbles are seldom formed. It has been experimentally reported that sonoluminescence intensity increases by addition of an appropriate amount of solutes because the number of active bubbles increases [86, 87]. It has also been experimentally confirmed that bubble size is smaller in an aqueous surfactant solution compared to that in pure water in acoustic cavitation [88].

2.18 Bubble–Bubble Interaction

As discussed in Sect. 2.16, a pulsating bubble radiates an acoustic wave into surrounding liquid according to Eq. (2.101).

$$p = \frac{\rho}{r}\left(2R\dot{R}^{2} + R^{2}\ddot{R}\right) \tag{2.104}$$

where p is the acoustic pressure radiated from a bubble, ρ is the liquid density, and r is the distance from a radiating bubble. The influence of an acoustic wave radiated by the surrounding bubbles on bubble pulsation is called *bubble–bubble interaction* (Fig. 2.35) [33]. The effect is taken into account in the Keller equation (Eq. 2.31) simply by using the following $p_{\text{s}}(t)$.

$$p_{\text{s}}(t) = -A \sin \omega t + \rho \sum_{i}\frac{1}{r_i}\left(2R_i\dot{R}_i^{2} + R_i^{2}\ddot{R}_i\right) \tag{2.105}$$

where r_i is the distance from the bubble numbered i, R_i is the instantaneous radius of bubble numbered i, and summation is for all the surrounding bubbles. By neglecting terms of order $(R_i/r_i)(\dot{R}_j/c_\infty)$, the following equation is obtained from the Keller equation [75].

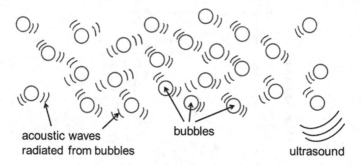

acoustic waves
radiated from bubbles

bubbles

ultrasound

Fig. 2.35 Bubble–bubble interaction. Pulsation of a bubble is influenced by the acoustic waves radiated by the surrounding bubbles. Reprinted with permission from Yasui [33]. Copyright (2016), Springer

$$\left(1 - \frac{\dot{R}}{c_\infty}\right) R\ddot{R} + \frac{3}{2}\dot{R}^2\left(1 - \frac{\dot{R}}{3c_\infty}\right) = \frac{1}{\rho_{L,\infty}}\left(1 + \frac{\dot{R}}{c_\infty}\right)(p_B + A\sin\omega t - p_\infty)$$
$$+ \frac{R}{c_\infty\rho_{L,\infty}}\frac{dp_B}{dt} - \sum_i \frac{1}{r_i}\left(2R_i\dot{R}_i^2 + R_i^2\ddot{R}_i\right)$$

$$(2.106)$$

In order to numerically solve Eq. (2.106), the number of equations necessary to simultaneously solve is equivalent to the number of bubbles. However, it is computationally expensive. Thus, the number of equations is dramatically reduced to 1 by assuming that the radius of each bubble is the same for all the bubbles. Furthermore, the spatial distribution of the bubbles is assumed to be uniform. Under this homogeneous bubble approximation, a set of Eqs. (2.106) is reduced to a single Eq. (2.107) [74, 89–91].

$$\left(1 - \frac{\dot{R}}{c_\infty}\right) R\ddot{R} + \frac{3}{2}\dot{R}^2\left(1 - \frac{\dot{R}}{3c_\infty}\right) = \frac{1}{\rho_{L,\infty}}\left(1 + \frac{\dot{R}}{c_\infty}\right)(p_B + A\sin\omega t - p_\infty)$$
$$+ \frac{R}{c_\infty\rho_{L,\infty}}\frac{dp_B}{dt} - S\left(2R\dot{R}^2 + R^2\ddot{R}\right)$$

$$(2.107)$$

where

$$S = \sum_i \frac{1}{r_i} = \int_{l_{min}}^{l_{max}} \frac{4\pi r^2 n}{r}\,dr = 2\pi n\left(l_{max}^2 - l_{min}^2\right) \approx 2\pi n l_{max}^2 \qquad (2.108)$$

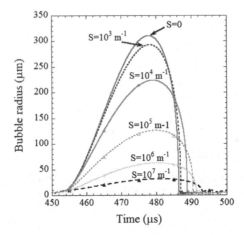

Fig. 2.36 Results of the numerical simulations on radius–time curves for various coupling strengths (S) of bubble–bubble interaction. The frequency and pressure amplitude of ultrasound are 20 kHz and 10 bar, respectively. The ambient static pressure is 5 atm. The ambient bubble radius is 5 μm. The liquid viscosity is 1 mPa s. Reprinted with permission from Yasui et al. [90]. Copyright (2011), AIP Publishing LLC

where l_{max} is the radius of a bubble cloud, l_{min} is the distance between a bubble and a nearest bubble, $l_{max} \gg l_{min}$ is assumed in the last equation, and n is the number density of the bubbles. The factor S is called the *coupling strength of bubble–bubble interaction* [74, 89–92]. The results of several numerical simulations of Eq. (2.107) have shown that bubble expansion is more strongly suppressed as the coupling strength increases (Fig. 2.36) [90].

Following the method described in Sect. 2.12, the resonance angular frequency of a bubble is derived from Eq. (2.107) as follows [89].

$$
\omega_0 = \sqrt{\frac{3\gamma p_\infty + (3\gamma - 1)2\sigma/R_0}{\rho_{L,\infty} R_0 \left(R_0 + SR_0^2 + 4\mu/c_\infty \rho_{L,\infty}\right)}}
\tag{2.109}
$$

Thus, the resonance frequency of a bubble considerably decreases as the coupling strength (S) increases above about 10^5 m^{-1} [89].

2.19 Acoustic Cavitation Noise

In experiments of acoustic cavitation, acoustic noise from cavitating liquid is often heard especially at relatively low driving ultrasonic frequencies. Such noise is called *acoustic cavitation noise*. From experimental measurement of acoustic emission originating from a single bubble in SBSL, most of the acoustic emissions from a pulsating bubble occur at the end of a violent bubble collapse (Fig. 2.37)

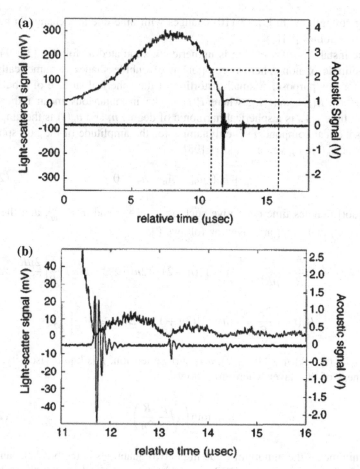

Fig. 2.37 Radius–time curve measured by light scattering and acoustic signal measured with a hydrophone for a single bubble in SBSL experiment. The ultrasonic frequency was 33.8 kHz. **a** For about 18 μs. **b** A detailed view of the boxed area in (**a**). Reprinted with permission from Matula et al. [93]. Copyright (1998), AIP Publishing LLC

[93]. In other words, most of acoustic cavitation noise originates in shock waves emitted from acoustic cavitation bubbles (Sect. 2.13).

According to Eq. (2.105), acoustic pressure (P) radiated from pulsating bubbles is given as follows.

$$P = \rho \sum_i \frac{1}{r_i} \left(2R_i\dot{R}_i^2 + R_i^2\ddot{R}_i\right) = S\rho\left(R^2\ddot{R} + 2R\dot{R}^2\right) \qquad (2.110)$$

where the homogeneous bubble approximation is used in the last equation. When some of bubbles disintegrate into "daughter" bubbles by shape instability, the coupling strength (S) in Eq. (2.110) changes with time due to a change in number density of bubbles [94].

Shape instability of a bubble is numerically simulated as follows [38, 94, 95]. The amplitude of non-spherical component of bubble shapes is numerically calculated for this purpose. A small distortion of the spherical surface of a bubble is described by $R(t) + a_n(t)Y_n$, where $R(t)$ is the instantaneous mean radius of a bubble at time t, Y_n is a spherical harmonic of degree n, and $a_n(t)$ is the amplitude of non-spherical component. The dynamics for the amplitude of the non-spherical component $a_n(t)$ is given as follows [95].

$$\ddot{a}_n + B_n(t)\dot{a}_n - A_n(t)a_n = 0 \tag{2.111}$$

where [dot] denotes time derivative such as $\ddot{a}_n = \frac{d^2 a_n}{dt^2}$ and $\dot{a}_n = \frac{da_n}{dt}$, and the coefficients $A_n(t)$ and $B_n(t)$ are given as follows [95].

$$A_n(t) = (n-1)\frac{\ddot{R}}{R} - \frac{\beta_n \sigma}{\rho R^3} - \left[(n-1)(n+2) + 2n(n+2)(n-1)\frac{\delta}{R} \right] \frac{2\mu \dot{R}}{R^3} \tag{2.112}$$

$$B_n(t) = \frac{3\dot{R}}{R} + \left[(n+2)(2n+1) - 2n(n+2)^2 \frac{\delta}{R} \right] \frac{2\mu}{R^2} \tag{2.113}$$

where $\beta_n = (n-1)(n+1)(n+2)$, σ is surface tension, μ is liquid viscosity, and δ is thickness of thin layer where fluid flows.

$$\delta = \min\left(\sqrt{\frac{\mu}{\omega}}, \frac{R}{2n} \right) \tag{2.114}$$

where min means the minimum value in the two quantities in the brackets, and ω is angular frequency of ultrasound. When $a_n(t)$ becomes larger than $R(t)$, then a bubble is considered to be disintegrated into "daughter" bubbles by the shape instability [95].

In actual experiments, the acoustic cavitation noise is measured with a hydrophone. A hydrophone has a cutoff frequency, and its response is approximately modeled using the following equation [94, 96].

$$\ddot{U} + 2\gamma_{dh}\pi f_c \dot{U} + 4\pi^2 f_c^2 U = P(t) - A \sin \omega t \tag{2.115}$$

where U is the hydrophone signal, γ_{dh} is the coefficient for damping, f_c is the characteristic frequency of hydrophone, $P(t)$ is the acoustic pressure radiated from pulsating bubbles given by Eq. (2.110), and $-A \sin \omega t$ is the acoustic pressure of the driving ultrasound. The characteristic frequency (f_c) of the hydrophone is related to the cutoff frequency since the sensitivity of hydrophone above f_c is lower than that below f_c. For a larger value of f_c, the intensity of the high-frequency component becomes stronger. For a larger value of γ_{dh}, the frequency cutoff becomes sharper.

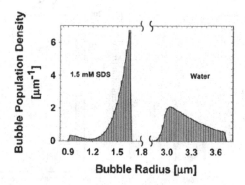

Fig. 2.38 Bubble population density versus the bubble radius for water and 1.5 mM SDS solution, estimated from the experimental data on quenching of SL intensity by increasing pulse-off time of pulsed ultrasound (515 kHz). Reprinted with permission from Lee et al. [88]. Copyright (2005), American Chemical Society

The frequency spectra of the acoustic cavitation noise measured with a hydrophone consist of the strongest peak at a driving ultrasonic frequency (515 kHz) (fundamental frequency), discrete peaks at integer multiple of the fundamental frequency (harmonics), discrete weaker peaks at a half of the fundamental frequency (subharmonic) and its integer multiple (ultra-harmonics), and weaker broadband continuum component (broadband noise) [97]. Surprisingly, the frequency spectra from low-concentration surfactant (sodium dodecyl sulfate, SDS) solution in the concentration range of 0.5–2 mM consist of only discrete peaks at fundamental frequency and its integer multiple [97]. The broadband component is significantly weaker than that in pure water. As already pointed out in Sect. 2.17, the bubble size is smaller in aqueous surfactant solution than that in pure water in a certain concentration range. For example, the bubble radii in 1.5 mM SDS aqueous solution were in the range of 0.9–1.7 μm, while those in pure water were in the range of 2.8–3.7 μm, which were experimentally measured from quenching of sonoluminescence intensity by increasing pulse-off time in pulsed ultrasound at 515 kHz (Fig. 2.38) [88].

According to the numerical simulations of amplitude of non-spherical component of bubble shape (Eq. 2.111) using the modified Keller equation (Eq. 2.107), a bubble is shape stable when the ambient bubble radius is 1.5 μm which is typical in a 1.5 mM SDS solution [94]. Thus, in this case, there is no temporal variation in the number of bubbles and the coupling strength. The bubble pulsation is temporally periodic (Fig. 2.39a), and the acoustic emission is also temporally periodic because there is no temporal variation in the number of bubbles (Fig. 2.39b) [94]. Accordingly, the hydrophone signal simulated using Eq. (2.115) is also temporally periodic (Fig. 2.39c). Then, the frequency spectrum of the hydrophone signal only consists of fundamental frequency (515 kHz) and its harmonics (Fig. 2.39d). It is consistent with the experimental measurement [97]. Even with strong shock waves emitted from bubbles (Fig. 2.39b), there is no broadband component in the acoustic

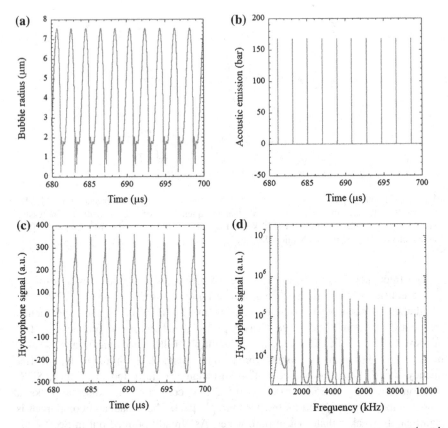

Fig. 2.39 Results from numerical simulations with a constant coupling strength of 10^4 m^{-1} (without a temporal fluctuation in the number of bubbles). The ambient bubble radius is 1.5 μm, which is typical in low-concentration SDS solutions. The frequency and pressure amplitude of ultrasound are 515 kHz and 2.6 bar, respectively. **a** Bubble radius. **b** Acoustic pressure radiated from bubbles (Eq. 2.110). **c** Hydrophone signal (Eq. 2.115). **d** Frequency spectrum of the hydrophone signal. Reprinted with permission from Yasui et al. [94]. Copyright (2010), Elsevier

cavitation spectrum. In other words, broadband noise is not solely originated in shock wave emissions.

In the case of ambient radius of 3 μm which is typical in pure water at 515 kHz, a bubble disintegrates into "daughter" bubbles in four (4) acoustic cycles [94]. Thus, in this case, there is a temporal fluctuation in the number of bubbles as well as the coupling strength. Even with this temporal variation, the radius–time curve is almost temporally periodic (Fig. 2.40a). However, the peaks in acoustic pressure due to shock wave emissions temporally fluctuate according to the temporal fluctuation in number of bubbles (Fig. 2.40b). As a result, small peaks in hydrophone signal caused by the shock waves also temporally fluctuate (Fig. 2.40c). Then, there is a strong broadband component in the frequency spectrum of the hydrophone signal (Fig. 2.40d) [94]. Thus, the origin of the broadband noise is temporal

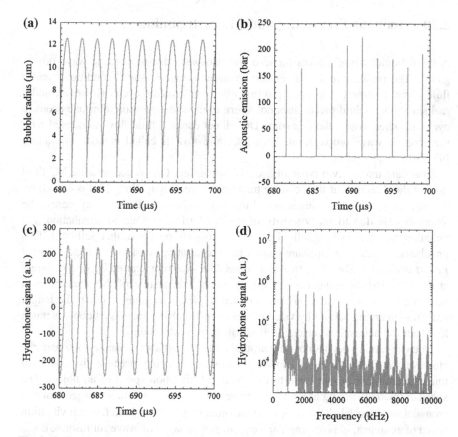

Fig. 2.40 Results from numerical simulations with a temporal fluctuation in the number of bubbles (coupling strength). The ambient bubble radius is 3 μm, which is typical in pure water. The frequency and pressure amplitude of ultrasound are the same as those shown in Fig. 2.39. **a** Bubble radius. **b** Acoustic pressure radiated from bubbles (Eq. 2.110). **c** Hydrophone signal (Eq. 2.115). **d** Frequency spectrum of the hydrophone signal. Reprinted with permission from Yasui et al. [94]. Copyright (2010), Elsevier

fluctuation in the number of bubbles. Broadband noise also results from non-periodic chaotic pulsation of bubbles as well as from initial transient pulsation of bubbles before reaching steady-state pulsation [94]. However, the contribution to the actual broadband noise is minor at least under the experimental condition of Ref. [97].

Subharmonic and ultra-harmonic originate from periodic bubble pulsation with doubled acoustic period of larger bubbles with ambient radius of 5 μm under this condition [94]. When the number of bubbles is much larger and the coupling strength is much larger, broadband noise as well as sub- and ultra-harmonics is also resulted from the bubble–bubble interaction [the last term in Eq. (2.107)] [94]. Further studies are required on the origin of the broadband noise [98].

2.20   Acoustic Streaming and Microstreaming

A fluid (liquid) parcel moves forward and backward due to acoustic wave propagation. The periodic motion is exactly symmetric under "ideal" conditions, and a fluid parcel returns to the same position after one acoustic cycle. However, under real situations, a fluid parcel does not return to the same position after one acoustic cycle. In other words, there is some DC (direct current) fluid flow associated with the acoustic wave propagation. Such DC fluid flow is called *acoustic streaming* [99–101].

There are mainly two types of acoustic streaming. One is accelerating DC fluid flow in the direction of the acoustic traveling wave propagation. This is caused by the attenuation of an acoustic traveling wave (ultrasound). In many cases, the attenuation is due to the viscosity of the fluid (liquid). Due to attenuation, the radiation pressure pushing a fluid parcel becomes stronger than that pulling it. This unbalance of radiation pressure causes the accelerating fluid flow which is called *Eckert streaming*. The other type is a vortex-like streaming caused by viscous stress at the boundary layer near a wall or an object irrespective of the situation of a traveling or standing wave, which is called *Rayleigh streaming*. When the length scale for streaming caused by viscous stress near an object such as a bubble is much less than the acoustic wavelength, it is called *microstreaming*.

Microstreaming occurs not only around a bubble but also around a solid particle. However, it is much more significant around a bubble because the speed of microstreaming is proportional to the square of vibration speed of an object. The speed of microstreaming around a pulsating bubble is 10^2–10^6 times larger than that around a solid particle because it is on the order of $U^2/\omega a$, where U is the vibration speed of an object, ω is the angular frequency of an acoustic wave (ultrasound), and a is radius of an object [102]. Thus, the term "microstreaming" is usually used for liquid streaming around a pulsating bubble. In Fig. 2.41, some examples of pattern of microstreaming around a pulsating bubble on a solid surface are shown [103].

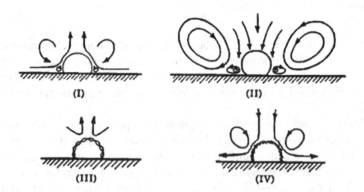

Fig. 2.41 Four types of microstreaming around a pulsating bubble. Reprinted with permission from Elder [103]. Copyright (1959), AIP Publishing LLC

Liquid flow associated with acoustic cavitation is highly complex. Ultrasound is strongly attenuated by cavitation bubbles, which intensifies Eckert streaming. Bubbles move due to primary and secondary Bjerknes forces, and liquid flow is influenced by the drag forces of bubbles. There is also microstreaming around pulsating bubbles. Rayleigh streaming also occurs near a wall. Liquid flow associated with acoustic cavitation is not fully understood at present [104].

References

1. Atkins P, de Paula J (2010) Atkins' physical chemistry, 9th edn. Oxford University Press, Oxford
2. Yasui K (2015) Dynamics of acoustic bubbles. In: Grieser F, Choi PK, Enomoto N, Harada H, Okitsu K, Yasui K (eds) Sonochemistry and the acoustic bubble. Elsevier, Amsterdam
3. Yasui K (2016) Mechanism for stability of ultrafine bubbles. Jpn J Multiph Flow 30:19–26 (in Japanese)
4. Keller JB, Miksis M (1980) Bubble oscillations of large amplitude. J Acoust Soc Am 68:628–633. doi:10.1121/1.384720
5. Prosperetti A, Lezzi A (1986) Bubble dynamics in a compressible liquid. Part 1. First-order theory. J Fluid Mech 168:457–478. doi:10.1017/S0022112086000460
6. Yasui K (1996) Variation of liquid temperature at bubble wall near the sonoluminescence threshold. J Phys Soc Jpn 65:2830–2840. doi:10.1143/JPSJ.65.2830
7. Pozrikidis C (2017) Fluid dynamics, 3rd edn. Springer, New York
8. Storey BD, Szeri AJ (1999) Mixture segregation within sonoluminescence bubbles. J Fluid Mech 396:203–221. doi:10.1017/S0022112099005984
9. Yasui K (2001) Effect of liquid temperature on sonoluminescence. Phys Rev E 64:016310. doi:10.1103/PhysRevE.64.016310
10. Kundu PK (1990) Fluid mechanics. Academic Press, San Diego
11. Akhatov I, Lindau O, Topolnikov A, Mettin R, Vakhitova N, Lauterborn W (2001) Collapse and rebound of a laser-induced cavitation bubble. Phys Fluids 13:2805–2819. doi:10.1063/1.1401810
12. Muller S, Bachmann M, Kroninger D, Kurz T, Helluy P (2009) Comparison and validation of compressible flow simulations of laser-induced cavitation bubbles. Comput Fluid 38:1850–1862. doi:10.1016/j.compfluid.2009.04.004
13. Gould H, Tobochnik J, Christian W (2007) An introduction to computer simulation methods, applications to physical systems, 3rd edn. Pearson, Addison Wesley, San Francisco
14. Yasui K (1997) Alternative model of single-bubble sonoluminescence. Phys Rev E 56:6750–6760. doi:10.1103/PhysRevE.56.6750
15. Yasui K (1996) A new formulation of bubble dynamics for sonoluminescence. Ph.D. thesis, Waseda University, Japan
16. Yasui K, Tuziuti T, Kanematsu W (2016) Extreme conditions in a dissolving air nanobubble. Phys Rev E 94:013106. doi:10.1103/PhysRevE.94.013106
17. Yasui K, Tuziuti T, Sivakumar M, Iida Y (2005) Theoretical study of single-bubble sonochemistry. J Chem Phys 122:224706. doi:10.1063/1.1925607
18. Toegel R, Lohse D (2003) Phase diagrams for sonoluminescing bubbles: a comparison between experiment and theory. J Chem Phys 118:1863–1875. doi:10.1063/1.1531610
19. Storey BD, Szeri AJ (2001) A reduced model of cavitation physics for use in sonochemistry. Proc R Soc Lond A 457:1685–1700. doi:10.1098/rspa.2001.0784

20. Lohse D, Brenner MP, Dupont TF, Hilgenfeldt S, Johnston B (1997) Sonoluminescing air bubbles rectify argon. Phys Rev Lett 78:1359–1362. doi:10.1103/PhysRevLett.78.1359

21. Brenner MP, Hilgenfeldt S, Lohse D (2002) Single-bubble sonoluminescence. Rev Mod Phys 74:425–484. doi:10.1103/RevModPhys.74.425

22. Yasui K, Tuziuti T, Sivakumar M, Iida Y (2004) Sonoluminescence. Appl Spectrosc Rev 39:399–436. doi:10.1081/ASR-200030202

23. Storey BD, Szeri AJ (2000) Water vapour, sonoluminescence and sonochemistry. Proc R Soc Lond A 456:1685–1709. doi:10.1098/rspa.2000.0582

24. Yasui K (2001) Single-bubble sonoluminescence from noble gases. Phys Rev E 63:035301. doi:10.1103/PhysRevE.63.035301

25. Yasui K (2002) Segregation of vapor and gas in a sonoluminescing bubble. Ultrasonics 40:643–647. doi:10.1016/S0041-624X(02)00190-7

26. Schrage RW (1953) A theoretical study of interphase mass transfer. Columbia University Press, New York

27. Fujikawa S, Akamatsu T (1980) Effects of the no-equilibrium condensation of vapour on the pressure wave produced by the collapse of a bubble in a liquid. J Fluid Mech 97:481–512. doi:10.1017/S0022112080002662

28. Matsumoto M (1996) Molecular dynamics simulation of interphase transport at liquid surfaces. Fluid Phase Equilib 125:195–203. doi:10.1016/S0378-3812(96)03123-8

29. Yasui K (1998) Effect of non-equilibrium evaporation and condensation on bubble dynamics near the sonoluminescence threshold. Ultrasonics 36:575–580. doi:10.1016/S0041-6244(97)00107-8

30. Suslick KS, Hammerton DA, Cline RE Jr (1986) The sonochemical hot spot. J Am Chem Soc 108:5641–5642. doi:10.1021/ja00278a055

31. Hua I, Hochemer RH, Hoffmann MR (1995) Sonolytic hydrolysis of p-nitrophenyl acetate: the role of supercritical water. J Phys Chem 99:2335–2342. doi:10.1021/j100008a015

32. Moriwaki H, Takagi Y, Tanaka M, Tsuruho K, Okitsu K, Maeda Y (2005) Sonochemical decomposition of perfluorooctane sulfonate and perfluorooctanoic acid. Environ Sci Technol 39:3388–3392. doi:10.1021/es040342v

33. Yasui K (2016) Unsolved problems in acoustic cavitation. In: Ashokkumar M, Cavalieri F, Chemat F, Okitsu K, Sambandam A, Yasui K, Zisu B (eds) Handbook of ultrasonics and sonochemistry. Springer, Singapore

34. Vuong VQ, Szeri AJ (1996) Sonoluminescence and diffusive transport. Phys Fluids 8:2354–2364. doi:10.1063/1.869020

35. Kamath V, Prosperetti A, Egolfopoulos FN (1993) A theoretical study of sonoluminescence. J Acoust Soc Am 94:248–260. doi:10.1121/1.407083

36. Shen Y, Yasui K, Sun Z, Mei B, You M, Zhu T (2016) Study on the spatial distribution of the liquid temperature near a cavitation bubble wall. Ultrason Sonochem 29:394–400. doi:10.1016/j.ultsonch.2015.10.015

37. Eller A, Flynn HG (1965) Rectified diffusion during nonlinear pulsations of cavitation bubbles. J Acoust Soc Am 37:493–503. doi:10.1121/1.1909357

38. Yasui K (2002) Influence of ultrasonic frequency on multibubble sonoluminescence. J Acoust Soc Am 112:1405–1413. doi:10.1121/1.1502898

39. Leighton TG (1994) The acoustic bubble. Academic Press, London

40. Leong T, Ashokkumar M, Kentish S (2016) The growth of bubbles in an acoustic field by rectified diffusion. In: Ashokkumar M, Cavalieri F, Chemat F, Okitsu K, Sambandam A, Yasui K, Zisu B (eds) Handbook of ultrasonics and sonochemistry. Springer, Singapore

41. Crum LA (1980) Measurements of the growth of air bubbles by rectified diffusion. J Acoust Soc Am 68:203–211. doi:10.1121/1.384624

42. Louisnard O, Gomez F (2003) Growth by rectified diffusion of strongly acoustically forced gas bubbles in nearly saturated liquids. Phys Rev E 67:036610. doi:10.1103/PhysRevE.67.036610

43. Yasui K, Tuziuti T, Lee J, Kozuka T, Towata A, Iida Y (2008) The range of ambient radius for an active bubble in sonoluminescence and sonochemical reactions. J Chem Phys 128:184705. doi:10.1063/1.2919119

44. Yasui K (1997) Chemical reactions in a sonoluminescing bubble. J Phys Soc Jpn 66:2911–2920. doi:10.1143/JPSJ.66.2911

45. Yasui K, Tuziuti T, Iida Y, Mitome H (2003) Theoretical study of the ambient-pressure dependence of sonochemical reactions. J Chem Phys 119:346–356. doi:10.1063/1.1576375

46. Didenko YT, Suslick KS (2002) The energy efficiency of formation of phtons, radicals and ions during single-bubble cavitation. Nature (London) 418:394–397. doi:10.1038/nature00895

47. Matula TJ, Crum LA (1998) Evidence for gas exchange in single-bubble sonoluminescence. Phys Rev Lett 80:865–868. doi:10.1103/PhysRevLett.80.865

48. Yasui K, Tuziuti T, Kozuka T, Towata A, Iida Y (2007) Relationship between the bubble temperature and main oxidant created inside an air bubble under ultrasound. J Chem Phys 127:154502. doi:10.1063/1.2790420

49. Yasui K, Tuziuti T, Iida Y (2004) Optimum bubble temperature for the sonochemical production of oxidants. Ultrasonics 42:579–584. doi:10.1016/j.ultras.2003.12.005

50. Hart EJ, Henglein A (1985) Free radical and free atom reactions in the sonolysis of aqueous iodide and formate solutions. J Phys Chem 89:4342–4347. doi:10.1021/j100266a038

51. Yasui K (2002) Effect of volatile solutes on sonoluminescence. J Chem Phys 116:2945–2954. doi:10.1063/1.1436122

52. Ashokkumar M, Crum LA, Frensley CA, Grieser F, Matula TJ, McNamara WB III, Suslick KS (2000) Effect of solutes on single-bubble sonoluminescence in water. J Phys Chem 104:8462–8465. doi:10.1021/jp000463r

53. Guan J, Matula TJ (2003) Time scales for quenching single-bubble sonoluminescence in the presence of alcohols. J Phys Chem B 107:8917–8921. doi:10.1021/jp026494z

54. Kinsler LE, Frey AR, Coppens AB, Sanders JV (1982) Fundamentals of acoustics, 3rd edn. Wiley, New York

55. Landau LD, Lifshitz EM (1987) Fluid mechanics, 2nd edn. (trans: Sykes JB, Reid WH). Elsevier, Amsterdam

56. Holzfuss J, Ruggeberg M, Billo A (1998) Shock wave emissions of a sonoluminescing bubble. Phys Rev Lett 81:5434–5437. doi:10.1103/PhysRevLett.81.5434

57. Hickling R, Plesset MS (1964) Collapse and rebound of a spherical bubble in water. Phys Fluids 7:7–14. doi:10.1063/1.1711058

58. Wu CC, Roberts PH (1993) Shock-wave propagation in a sonoluminescing gas bubble. Phys Rev Lett 70:3424–3427. doi:10.1103/PhysRevLett.70.3424

59. Moss WC, Clarke DB, White JW, Young DA (1994) Hydrodynamic simulations of bubble collapse and picosecond sonoluminescence. Phys Fluids 6:2979–2985. doi:10.1063/1.868124

60. Nigmatulin RI, Akhatov IS, Topolnikov AS, Bolotnova RK, Vakhitova NK, Lahey RT Jr, Taleyarkhan RP (2005) Theory of supercompression of vapor bubbles and nanoscale thermonuclear fusion. Phys Fluids 17:107106. doi:10.1063/1.2104556

61. Yuan L, Cheng HY, Chu MC, Leung PT (1998) Physical parameters affecting sonoluminescence: a self-consistent hydrodynamic study. Phys Rev E 57:4265–4280. doi:10.1103/PhysRevE.57.4265

62. Cheng HY, Chu MC, Leung PT, Yuan L (1998) How important are shock waves to single-bubble sonoluminescence? Phys Rev E 58:R2705–R2708. doi:10.1103/PhysRevE.58.R2705

63. Yuan L (2005) Sonochemical effects on single-bubble sonoluminescence. Phys Rev E 72:046309. doi:10.1103/PhysRevE.72.046309

64. An Y (2006) Mechanism of single-bubble sonoluminescence. Phys Rev E 74:026304. doi:10.1103/PhysRevE.74.026304

65. An Y, Li C (2008) Spectral lines of OH radicals and Na atoms in sonoluminescence. Phys Rev E 78:046313. doi:10.1103/PhysRevE.78.046313

66. Vuong VQ, Szeri AJ, Young DA (1999) Shock formation within sonoluminescence bubbles. Phys Fluids 11:10–17. doi:10.1063/1.869920

67. Ohl CD, Arora M, Dijkink R, Janve V, Lohse D (2006) Surface cleaning from laser-induced cavitation bubbles. Appl Phys Lett 89:074102. doi:10.1063/1.2337506

68. Plesset MS, Chapman RB (1971) Collapse of an initially spherical vapour cavity in the neighbourhood of a solid boundary. J Fluid Mech 47:283–290. doi:10.1017/S002211207 1001058

69. Lamminen MO, Walker HW, Weavers LK (2004) Mechanism and factors influencing the ultrasonic cleaning of particle-fouled ceramic membranes. J Membr Sci 237:213–223. doi:10.1016/j.memsci.2004.02.031

70. Bremond N, Arora M, Dammer SM, Lohse D (2006) Interaction of cavitation bubbles on a wall. Phys Fluids 18:121505. doi:10.1063/1.2396922

71. Calvisi ML, Lindau O, Blake JR, Szeri AJ (2007) Shape stability and violent collapse of microbubbles in acoustic traveling waves. Phys Fluids 19:047101. doi:10.1063/1.2716633

72. Mettin R (2005) Bubble structures in acoustic cavitation. In: Doinikov AA (ed) Bubble and particle dynamics in acoustic fields: modern trends and applications. Research Signpost, Trivandrum

73. Matula TJ, Cordry SM, Roy RA, Crum LA (1997) Bjerknes force and bubble levitation under single-bubble sonoluminescence conditions. J Acoust Soc Am 102:1522–1527. doi:10.1121/1.420065

74. Yasui K, Iida Y, Tuziuti T, Kozuka T, Towata A (2008) Strongly interacting bubbles under an ultrasonic horn. Phys Rev E 77:016609. doi:10.1103/PhysRevE.77.016609

75. Mettin R, Akhatov I, Parlitz U, Ohl CD, Lauterborn W (1997) Bjerknes forces between small cavitation bubbles in a strong acoustic field. Phys Rev E 56:2924–2931. doi:10.1103/PhysRevE.56.2924

76. Firouzi M, Howes T, Nguyen AV (2015) A quantitative review of the transition salt concentration for inhibiting bubble coalescence. Adv Colloid Interface Sci 222:305–318. doi:10.1016/j.cis.2014.07.005

77. Prince MJ, Blanch HW (1990) Transition electrolyte concentrations for bubble coalescence. AIChE J 36:1425–1429. doi:10.1002/aic.690360915

78. Prince MJ, Blanch HW (1990) Bubble coalescence and break-up in air-sparged bubble columns. AIChE J 36:1485–1499. doi:10.1002/aic.690361004

79. Craig VSJ, Ninham BW, Pashley RM (1993) Effect of electrolytes on bubble coalescence. Nature (London) 364:317–319. doi:10.1038/364317a0

80. Christenson HK, Yaminsky VV (1995) Solute effects on bubble coalescence. J Phys Chem 99:10420. doi:10.1021/j100025a052

81. Oolman TO, Blanch HW (1986) Bubble coalescence in stagnant liquids. Chem Eng Commun 43:237–261. doi:10.1080/00986448608911334

82. Lee JC, Meyrick DL (1970) Gas-liquid interfacial areas in salt solutions in an agitated tank. Trans Inst Chem Eng 48:T37–T45

83. Marrucci G (1969) A theory of coalescence. Chem Eng Sci 24:975–985. doi:10.1016/0009-2509(69)87006-5

84. Iida Y, Ashokkumar M, Tuziuti T, Kozuka T, Yasui K, Towata A, Lee J (2010) Bubble population phenomena in sonochemical reactor: I estimation of bubble size distribution and its number density with pulsed sonication—laser diffraction method. Ultrason Sonochem 17:473–479. doi:10.1016/j.ultsonch.2009.08.018

85. Iida Y, Ashokkumar M, Tuziuti T, Kozuka T, Yasui K, Towata A, Lee J (2010) Bubble population phenomena in sonochemical reactor: II. Estimation of bubble size distribution and its number density by simple coalescence model calculation. Ultrason Sonochem 17:480–486. doi:10.1016/j.ultsonch.2009.08.017

86. Ashokkumar M, Hall R, Mulvaney P, Grieser F (1997) Sonoluminescence from aqueous alcohol and surfactant solutions. J Phys Chem 101:10845–10850. doi:10.1021/jp972477b

87. Sunartio D, Ashokkumar M, Grieser F (2005) The influence of acoustic power on multibubble sonoluminescence in aqueous solution containing organic solutes. J Phys Chem B 109:20044–20050. doi:10.1021/jp052747n

88. Lee J, Ashokkumar M, Kentish S, Grieser F (2005) Determination of the size distribution of sonoluminescence bubbles in a pulsed acoustic field. J Am Chem Soc 127:16810–16811. doi:10.1021/ja0566432

89. Yasui K, Lee J, Tuziuti T, Towata A, Kozuka T, Iida Y (2009) Influence of the bubble-bubble interaction on destruction of encapsulated microbubbles under ultrasound. J Acoust Soc Am 126:973–982. doi:10.1121/1.3179677

90. Yasui K, Towata A, Tuziuti T, Kozuka T, Kato K (2011) Effect of static pressure on acoustic energy radiated by cavitation bubbles in viscous liquids under ultrasound. J Acoust Soc Am 130:3233–3242. doi:10.1121/1.3626130

91. Yasui K, Kato K (2012) Bubble dynamics and sonoluminescence from helium or xenon in mercury and water. Phys Rev E 86:036320. doi:10.1103/PhysRevE.86.036320. Erratum. Phys Rev E 86:069901. doi:10.1103/PhysRevE.86.069901

92. Guedra M, Cornu C, Inserra C (2017) A derivation of the stable cavitation threshold accounting for bubble-bubble interactions. Ultrason Sonochem 38:168–173. doi:10.1016/j.ultsonch.2017.03.010

93. Matula TJ, Hallaj IM, Cleveland RO, Crum LA, Moss WC, Roy RA (1998) The acoustic emissions from single-bubble sonoluminescence. J Acoust Soc Am 103:1377–1382. doi:10.1121/1.421279

94. Yasui K, Tuziuti T, Lee J, Kozuka T, Towata A, Iida Y (2010) Numerical simulations of acoustic cavitation noise with the temporal fluctuation in the number of bubbles. Ultrason Sonochem 17:460–472. doi:10.1016/j.ultsonch.2009.08.014

95. Hilgenfeldt S, Lohse D, Brenner MP (1996) Phase diagrams for sonoluminescing bubbles. Phys Fluids 8:2808–2826. doi:10.1063/1.869131

96. Luther S, Sushchik M, Parlitz U, Akhatov I, Lauterborn W (2000) Is cavitation noise governed by a low-dimensional chaotic attractor? AIP Conf Proc 524:355–358

97. Ashokkumar M, Hodnett M, Zeqiri B, Grieser F, Price GJ (2007) Acoustic emission spectra from 515 kHz cavitation in aqueous solutions containing surface-active solutes. J Am Chem Soc 129:2250–2258. doi:10.1021/ja067960r

98. Lauterborn W, Mettin R (2015) Acoustic cavitation: bubble dynamics in high-power ultrasonic fields. In: Gallego-Juarez JA, Graff KF (eds) Power ultrasonics—applications of high-intensity ultrasound. Woodhead Publishing, Cambridge (Elsevier, Amsterdam)

99. Manasseh R (2016) Acoustic bubbles, acoustic streaming, and cavitation microstreaming. In: Ashokkumar M, Cavalieri F, Chemat F, Okitsu K, Sambandam A, Yasui K, Zisu B (eds) Handbook of ultrasonics and sonochemistry. Springer, Singapore

100. Beyer RT (1997) Nonlinear acoustics. Acoustical Society of America, New York

101. Yasui K, Izu N (2017) Effect of evaporation and condensation on a thermoacoustic engine: a Lagrangian simulation approach. J Acoust Soc Am 141:4398–4407. doi:10.1121/1.4985385

102. Nyborg WL (1958) Acoustic streaming near a boundary. J Acoust Soc Am 30:329–339. doi:10.1121/1.1909587

103. Elder SA (1959) Cavitation microstreaming. J Acoust Soc Am 31:54–64. doi:10.1121/1.1907611

104. Mettin R, Cairos C (2016) Bubble dynamics and observations. In: Ashokkumar M, Cavalieri F, Chemat F, Okitsu K, Sambandam A, Yasui K, Zisu B (eds) Handbook of ultrasonics and sonochemistry. Springer, Singapore

Chapter 3
Unsolved Problems

Abstract Although acoustic cavitation and bubble dynamics have been studied for more than 100 years, this field is still very active and there are a variety of unsolved problems. The old problem on cavitation nuclei is now in the spotlight because of the mysteries of bulk nanobubbles. With regard to sonochemical products, ammonia (NH_3) and oxygen atom (O) have not yet been fully studied. At the final moment of bubble collapse, solidification of water may take place near the bubble wall by the high pressure. A related unsolved problem is the mechanism of sonocrystallization that crystal nucleation is accelerated by acoustic cavitation. Plasma formation inside a sonoluminescing bubble has been confirmed by the spectroscopic observation. Is there a hot plasma core formed by shock wave focusing at the center of a bubble? What is quantitative theory of ionization-potential lowering inside a bubble nearly at liquid density? Why is the vibrational population distribution of OH radicals strongly in non-equilibrium inside a bubble? What are the roles of pulsed ultrasound and liquid surface vibration in acoustic field in the liquid? Is there any effect of a magnetic field on bubble dynamics? Are the extreme conditions inside a dissolving bubble real?

Keywords Cavitation threshold · Bulk nanobubbles · Ultrafine bubbles Ammonia · Solidification · A hot plasma core · Ionization-potential lowering Sonocrystallization · Non-equilibrium plasma · Magnetic field

3.1 Cavitation Nuclei (Bulk Nanobubbles)

Theoretical calculations of the tensile strength of pure water give values on the order of 1000 atm [1, 2]. The minimum acoustic amplitude for cavitation to occur is called the cavitation threshold. The experimentally determined cavitation threshold is more than one order of magnitude lower than the theoretical tensile strength of pure water (Fig. 3.1) [3, 4]. Furthermore, the cavitation threshold decreases as gas concentration in liquid increases. In highly degassed water, the cavitation threshold is about 80 atm [4]. In air-saturated water, the cavitation threshold is only about

© The Author(s) 2018
K. Yasui, *Acoustic Cavitation and Bubble Dynamics*,
Ultrasound and Sonochemistry, https://doi.org/10.1007/978-3-319-68237-2_3

Fig. 3.1 Cavitation threshold as a function of air concentration in pure water (degree of saturation) at an ultrasound frequency of 26.3 kHz [4]. Reprinted with permission from Yasui [3]. Copyright (2015), Elsevier

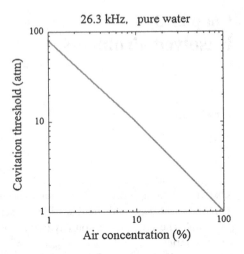

1 atm [4]. This is an experimental evidence that cavitation bubbles not only consist of water vapor but also gas (air). It should be noted, however, that the experimental data shown in Fig. 3.1 are only an example because cavitation threshold strongly depends on the concentration of impurities in water as well as that of crevices on the wall of a liquid container.

The discrepancy between theoretical tensile strength and the experimental cavitation threshold is due to the presence of cavitation nuclei in an actual liquid. There are mainly two types in cavitation nuclei: One is solid particles, and the other is tiny bubbles. From a crevice on a solid particle (or solid wall of a liquid container), bubbles are easily nucleated (Fig. 3.2) [3]. The interface of gas trapped inside a crevice is concave (Fig. 3.2 left). As a result, surface tension makes the internal gas pressure lower than the liquid pressure due to the Laplace pressure. Then, gas dissolution into the surrounding liquid is strongly retarded in contrast to the case of a normal bubble with convex interface. Under ultrasound, gas in a crevice expands during the rarefaction phase, and the gas pressure further decreases, causing the diffusion of gas dissolved in the surrounding liquid into gas trapped in a crevice. As a result, volume of gas in a crevice gradually increases. Finally, a new bubble is launched from a crevice due to buoyancy and radiation forces. This process is repeated because some gas is left in a crevice after a new bubble is pinched off.

Clean tiny bubbles immediately dissolve into liquid because the gas pressure inside a tiny bubble is considerably higher than the pressure of gas dissolved in liquid due to the Laplace pressure (Sect. 2.1). The time for the complete dissolution of a bubble into gas-saturated liquid is calculated by the Epstein–Plesset theory [Eq. (3.1)] (Fig. 3.3) [3, 5]:

$$t_{\text{diss}} = \frac{1}{3}\left(1 + \frac{R_0 p_\infty}{2\sigma}\right)\left(\frac{R_0^2 M_{\text{gas}} p_\infty}{D_{\text{gas}} c_{s0} R_g T}\right) \tag{3.1}$$

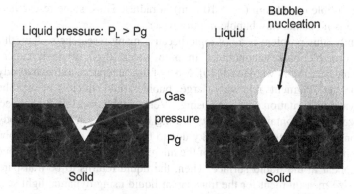

Fig. 3.2 Mechanism for bubble nucleation at a solid (particle) surface. Reprinted with permission from Yasui [3]. Copyright (2015), Elsevier

Fig. 3.3 Time for complete dissolution of an air bubble in water saturated with air as a function of the bubble radius calculated by Epstein–Plesset theory [Eq. (3.1)]. Reprinted with permission from Yasui [3]. Copyright (2015), Elsevier

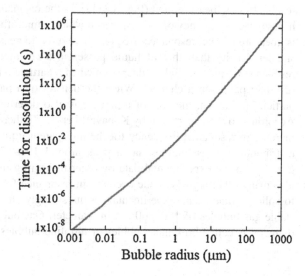

where t_{diss} is the time for the complete dissolution of a bubble into the gas-saturated liquid, R_0 is the initial bubble radius, p_∞ is the ambient liquid pressure, σ is the surface tension, M_{gas} is the molar weight of the gas, D_{gas} is the diffusion coefficient of gas in the liquid, c_{s0} is the saturated gas concentration in the liquid far from a bubble, R_g is the universal gas constant, and T is the temperature. Detailed derivation of Eq. (3.1) is given in Ref. [3]. In Fig. 3.3, the following quantities are used for an air bubble in air-saturated water at 20 °C: $p_\infty = 1$ atm $= 1.01325 \times 10^5$ Pa, $\sigma = 7.275 \times 10^{-2}$ N/m, $M_{gas} = 28.96 \times 10^{-3}$ kg/mol, $D_{gas} = 2.4 \times 10^{-9}$ m²/s, $c_{s0} = 2.3 \times 10^{-2}$ kg/m³, $R_g = 8.3145$ J/mol K, and $T = 293$ K. An air bubble of 1 μm ($= 1 \times 10^{-6}$ m) in radius completely dissolves into water saturated with air in 10 ms ($= 1 \times 10^{-2}$ s). It takes only 80 μs ($= 8 \times 10^{-5}$ s) for the complete dissolution of an air bubble of 100 nm ($= 1 \times 10^{-7}$ m) in radius. Thus, some researchers think that there is no stable tiny bubble in pure water.

Recently, this problem is a hot topic because many researchers claim that there are many stable bulk nanobubbles in pure water [6–9]. In their experiments, nanobubble generators were used [10]. Most of the generators utilize hydrodynamic cavitation to fragment relatively large bubbles into micro- or nanobubbles. Hydrodynamic cavitation occurs by using a Venturi tube, swirling flow, injection of pressurized water containing gas, etc. During the generation using hydrodynamic cavitation, liquid water becomes milky due to the presence of many microbubbles. After stopping the generation, most of the microbubbles move upward by buoyancy and disappear at the liquid surface. Then, the liquid returns to be transparent. By particle size measurement for the transparent liquid using dynamic light scattering, laser diffraction scattering, or particle tracking analysis, a peak in the size distribution of particles is observed in the range of 100–200 nm in diameter [8].

Many of the particles are confirmed to be bubbles rather than solid particles by resonant mass measurement (Fig. 3.4) [7]. In the resonant mass measurement, shift in resonance frequency of a cantilever in which a microfluidic channel is embedded is measured. The resonance frequency slightly increases when a particle with smaller density than that of liquid passes in a channel. On the other hand, the resonance frequency slightly decreases when a particle with larger density than that of liquid passes in a channel. When the density of a particle (gas bubble or dust particle) is known, the size of a particle is also determined by the measurement. According to the experiment by Kobayashi et al. [7], there were many signals of the increase in resonance frequency for the transparent liquid. The corresponding size distribution of gas bubbles has a peak around 100–150 nm in diameter, which agrees with the experimental data by other measuring methods [7, 8]. There were also many other signals of the decrease in resonance frequency, which correspond to solid particles. The experimental results strongly suggest that there exist many stable gas bubbles of 100–200 nm in diameter. Gas bubbles smaller than 1 μm in diameter is called bulk nanobubbles or ultrafine bubbles.

Fig. 3.4 Resonant mass measurement method

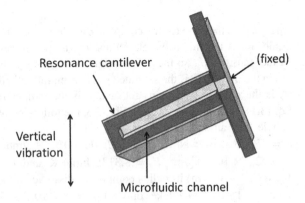

Resonance cantilever   (fixed)

Vertical
vibration

Microfluidic channel

What is the mechanism of stability for bulk nanobubbles? As already noted, a bubble of 100 nm in diameter should completely dissolve into liquid only in 80 μs. There are several models proposed to explain the stability of bulk nanobubbles. One is the skin model that a bubble is completely covered with organic materials or surfactants [11, 12]. The skin could prevent loss of gas by diffusion. The skin also reduces the Laplace pressure. However, in this model, a bubble should be 100% covered with organic materials or surfactants without any hole. It seems difficult, especially when organic material is solid-like.

Another is the model of the electrostatic negative (repulsive) pressure at the surface of a bubble (Fig. 3.5) [13]. The surface of a tiny bubble is negatively charged and has a zeta potential of about −40 mV [14–17]. In this case, the pressure inside a bubble is expressed as follows instead of Eq. (2.1) [13]:

$$p_{in} = p_B + \frac{2\sigma}{R} - \frac{\varepsilon\zeta^2}{R^2} \tag{3.2}$$

where ε is permittivity, and ζ is a zeta potential of a bubble. Numerical calculation shows, however, that the last term in Eq. (3.1) is negligible compared to the Laplace pressure if the zeta potential is kept as $\zeta = -40mV$ (Fig. 3.6) [13].

On a surface of a hydrophobic material, liquid water is repelled and a depletion layer is formed where the density of water is considerably reduced. The thickness of a depletion layer is 0.2–5 nm, and the density of liquid water is reduced to 44–94%, which was studied by neutron scattering and X-ray reflectivity measurement [18, 19]. In a depletion layer, gas dissolved in liquid water is trapped [20–23]. Thus, the concentration of gas becomes considerably higher on a surface of a hydrophobic

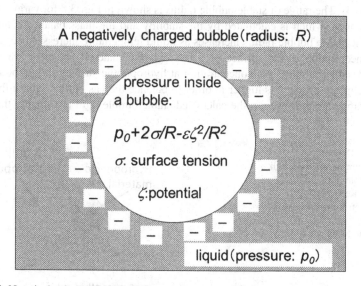

Fig. 3.5 Negatively charged bubble [13]

Fig. 3.6 Laplace pressure and absolute value of repulsive electrostatic pressure of a bubble as a function of bubble radius with logarithmic scale [13]

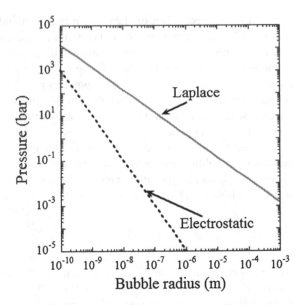

material than that away from a hydrophobic material. When a bubble is partly covered with a hydrophobic material, gas diffuses into a bubble near the peripheral edge of the hydrophobic material on the bubble surface (Fig. 3.7) [24]. When it balances with gas outflux from the other part of the uncovered bubble surface, dissolution of a bubble is stopped. When slight increase or decrease in the bubble radius results in decrease or increase in the initial radius, respectively, the mass balance is in an actually stable state. The model is called dynamic equilibrium model [24]. The range of stable bubble radius is shown in Fig. 3.8 for various liquid temperatures, which is calculated by the stability and the mass balance conditions [24]. The stable bubble radius increases as liquid temperature increases because the gas concentration at the surface of a hydrophobic material becomes lower. For smaller stable bubbles, gas concentration at hydrophobic surface should be higher as the Laplace pressure is higher in order to balance the gas outflux with influx. At room temperature, however, the calculated stable bubble radii are smaller than the

Fig. 3.7 Dynamic equilibrium model for a bulk nanobubble (an ultrafine bubble). Reprinted with permission from Yasui et al. [24]. Copyright (2016), American Chemical Society

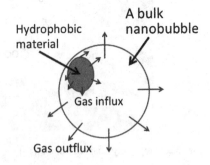

Fig. 3.8 Stable bubble radius as a function of the area covered with hydrophobic material for various temperatures according to the dynamic equilibrium model. Reprinted with permission from Yasui et al. [24]. Copyright (2016), American Chemical Society

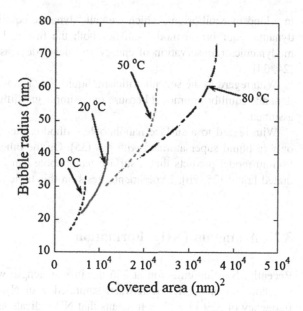

typical radii of bulk nanobubbles (ultrafine bubbles) of 50–100 nm (the diameter is in the range of 100–200 nm). There are mainly two possibilities for this discrepancy. One is that actual hydrophobic potential of a material covering bubble surface is lower than that assumed in the numerical calculations (1.7×10^{-20} J) [24–27]. Smaller hydrophobic potential results in lower gas concentration at hydrophobic surface and hence larger stable bubble as discussed above. The other is the aggregation of bulk nanobubbles as clusters in actual experiments. Actually, microbubbles were aggregated as clusters in aqueous surfactant solution under ultrasound [28].

The dynamic equilibrium model has been criticized due to the following two reasons. One is that the model may violate the laws of thermodynamics like a perpetual motion machine as there is a permanent circulating gas flow. The other is that liquid flow was not experimentally detected around a surface nanobubble [29, 30]. A surface nanobubble is a stable gas state with a lens shape on a solid surface, which has been confirmed both experimentally and theoretically [31]. The dynamic equilibrium model was first proposed for a surface nanobubble on a hydrophobic surface by Brenner and Lohse [32]. Gas concentration is significantly higher at a hydrophobic surface, and gas diffuses into a surface nanobubble near the contact line. The gas influx balances with outflux from the other part of interface of a surface nanobubble according to the dynamic equilibrium model.

In order to study on the first criticism, changes in energy and entropy for all the processes in the dynamic equilibrium for a bulk nanobubble are calculated in Ref. [24]. Then, it is shown that the total changes of energy and entropy are both zero for the entire process of the dynamic equilibrium. Total change of entropy is zero for an equilibrium state, while it should increase for the other cases [33]. Thus, the state is in a kind of equilibrium, which we call "dynamic equilibrium." It means that the dynamic equilibrium model satisfies both the first and the second laws of thermodynamics (conservation of energy and never decreasing entropy, respectively) [24, 34].

With regard to the second criticism, liquid flow is not necessarily required in the dynamic equilibrium model because only simple gas diffusion in quiescent liquid is assumed.

With regard to a surface nanobubble, orthodox theory predicts that it is stable only in liquid supersaturated with gas [35]. On the other hand, the dynamic equilibrium model predicts that a surface nanobubble can be stable even in undersaturated liquid [25, 26]. Experimental tests on this topic are required.

3.2 Ammonia (NH₃) Formation

Recently, NH-line emission at 336 nm in wavelength was observed in spectra of sonoluminescence (SL) from water saturated with N_2–Ar mixtures at ultrasonic frequency of 359 kHz [36]. It means that NH radicals are formed inside a bubble. On the other hand, at ultrasonic frequency of 20 kHz, NH-line was not observed [36]. For the both ultrasonic frequencies, OH-line was observed at about 310 nm. The reason for the dependence of SL spectra on ultrasonic frequency is still unclear.

There are a few experimental reports on the sonochemical formation of ammonia (NH_3) in liquid water in which N_2–Ar mixtures were dissolved [37, 38]. Main chemical products from bubbles are H_2, H_2O_2, NO_2^-, NO_3^-, and NH_3 (Fig. 3.9)

Fig. 3.9 Rate of the formation of various products as a function of the concentration of N_2 in the argon–nitrogen atmosphere. Ultrasonic frequency was 300 kHz, and insonation time was 40 min. Rates are overall rates. Reprinted with permission from Hart et al. [37]. Copyright (1986), American Chemical Society

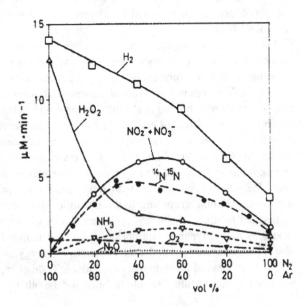

[37]. It has been suggested that NH$_3$ is formed inside a bubble through the following reactions [36, 38, 39]:

$$N_2 \rightarrow N + N \tag{3.3}$$

$$N + H \rightarrow NH \tag{3.4}$$

$$N_2 + H \rightarrow NH + N \tag{3.5}$$

$$NH + H_2 \rightarrow NH_3 \tag{3.6}$$

However, there have been no numerical simulations on NH$_3$ formation inside a bubble. Such simulations are required in future.

3.3 Solidification and Sonocrystallization

It has been theoretically predicted that the liquid pressure near the bubble wall increases to about 5 GPa at the end of the violent bubble collapse (the Rayleigh collapse) [40]. If the liquid temperature does not significantly increase near the bubble wall, the pressure and temperature near the bubble wall correspond to solid state (ice) (Fig. 3.10) [41]. Thus, it has been suggested that transient, high-pressure solidification of water occurs near the bubble wall at the end of violent collapse [41, 42].

Recently, in the experiment of the collapse of a single bubble produced by laser focusing in water irradiated with ultrasound under high static pressures up to 30 MPa, a spheroidal object was observed around a bubble after the violent collapse by using high-speed camera [43]. The ultrasonic frequency and the pressure amplitudes were 28 kHz and up to 35 MPa, respectively. The maximum radii of bubbles in the experiment ranged from approximately 0.8 to 1.8 mm, which were much larger than those in typical experiments of acoustic cavitation and sonochemistry [43]. The bubble wall speed at the violent collapse often exceeded 3000 m/s, and the maximum value at the most energetic collapses exceeded 7000 m/s, which was much higher than those in typical experiments of acoustic cavitation and sonochemistry. Observation of behaviors of the spheroidal objects formed around a bubble after the violent collapse suggests that a phase transition took place in the water near a bubble. It suggests that solidification of water occurred around a bubble. Further studies are required on this topic.

There are two other kinds of experiments on crystal nucleation accelerated by acoustic cavitation called sonocrystallization: One is ice crystallization in supercooled water [44–48], and the other is crystallization of a solute in supersaturated solution [49–51]. For the both cases, it has been experimentally known that bubbles play an important role in sonocrystallization. Furthermore, it has been

Fig. 3.10 Computed lines of adiabatic compression (dashed lines) of water superimposed on the equilibrium ice–water phase diagram. The Roman numerals in the phase diagram indicate different types of ice. Reprinted with permission from Hickling [41]. Copyright (1994), American Physical Society

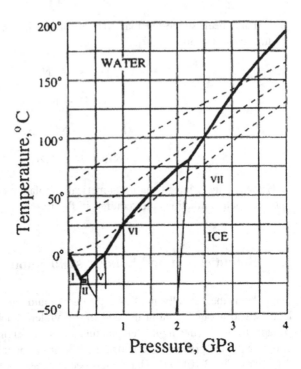

experimentally reported that the mean crystal size is reduced and that the size distribution of crystals becomes narrower [52, 53]. However, details of the mechanism of sonocrystallization are still unclear. For ice sonocrystallization, shock waves radiated by cavitation bubbles are believed to be important because solidification temperature increases as pressure increases [48]. On the other hand, it has been suggested that gas–liquid interface reduces the nucleation work of a solute and accelerates crystal nucleation [49, 54–56]. If this is the case, the violent bubble collapse is not necessary for sonocrystallization. There is a report, however, that transient cavitation (violent bubble collapse) is necessary for sonocrystallization [51]. Further studies are strongly required on the mechanism of sonocrystallization.

3.4   A Hot Plasma Core

When a spherical shock wave focuses at the center of a collapsing bubble, it has been theoretically predicted that a hot plasma core is formed at around the center of a bubble [57–59]. In such a theoretical model, the maximum temperature at the bubble center has been predicted to be as high as about 10 eV (10^5 K). Gases and vapor in a hot core should be highly ionized by the high temperature. Thus, it is called a hot plasma core.

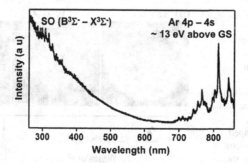

Fig. 3.11 MBSL spectrum from concentrated sulfuric acid under Ar. Sonication at 20 kHz (14 W/cm^2) with a Ti horn directly immersed in 95 wt% sulfuric acid at about 298 K. Reprinted with permission from Eddingsaas and Suslick [61]. Copyright (2007), American Chemical Society

Experimental observation of spectra of sonoluminescence in sulfuric acid suggested that a hot plasma core is indeed present inside a sonoluminescing bubble [60, 61]. In the spectrum, Ar lines originated from the transition between $4p$ and $4s$ states were observed (Fig. 3.11) [61]. The $4p$ state of Ar is about 13 eV above the ground state. On the other hand, the spectrum of Ar ($4p$–$4s$) emission is best fitted with an effective temperature of 8000 K (Fig. 3.12) [61]. This temperature of 8000 K is more than 1 order of magnitude lower than that required to excite Ar to $4p$ state from the ground one (13 eV). It suggests that excitation of Ar to $4p$ state is due to collisions with higher energy ions or electrons from a hot plasma core. In other words, the presence of a hot plasma core inside a sonoluminescing bubble is suggested. Recalling the discussion in Sect. 2.14, however, a shock wave is barely formed inside a collapsing bubble. Further studies are required on this topic. One possibility is that Ar at $4p$ state is formed by the recombination of Ar$^+$ ion and an electron. Ionization of gas molecules occurs much more frequently than excitation to a bound state because the density of states for ionized state is much higher than

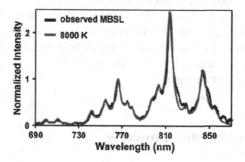

Fig. 3.12 Spectrum of Ar ($4p$–$4s$ manifold) emission from the MBSL compared to the best-fit synthetic spectrum, which gives an effective emission temperature of 8000 K. The synthetic spectra assumed thermal equilibration and a Lorentzian profile. Reprinted with permission from Eddingsaas and Suslick [61]. Copyright (2007), American Chemical Society

Fig. 3.13 MBSL of concentrated H_2SO_4 at different acoustic powers. **a** Photographs (10 s exposures) of different light-emitting regimes of MBSL of H_2SO_4, from left to right, with increasing acoustic intensity, filamentous, bulbous, and cone-shaped emission; **b** MBSL spectra of concentrated H_2SO_4 at the three acoustic intensities shown. As the acoustic power is increased, the Ar lines become weaker. Reprinted with permission from Eddingsaas and Suslick [61]. Copyright (2007), American Chemical Society

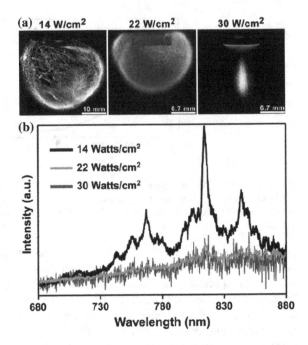

that of a bound state. According to statistical and quantum mechanics, probability of ionization or excitation is higher for larger density of final states [62, 63].

Another unsolved problem is disappearance of Ar ($4p$–$4s$) lines in multibubble sonoluminescence at increased intensity of ultrasound in sulfuric acid under an ultrasonic horn reported by Eddingsaas and Suslick (Fig. 3.13) [61]. Upon varying the acoustic power, large and abrupt changes in bubble-cloud dynamics, light intensity, and spectra were observed. There were three different light-emitting regimes as a function of acoustic intensity. At relatively low acoustic intensities, a wispy, filamentous emission was observed (Fig. 3.13). Above about 16 W/cm^2, the cavitating bubbles suddenly formed a bulb near the horn tip, creating a small globe of light consisting of very weak Ar lines along with the broad continuum. Above about 24 W/cm^2, light emission was observed from a cone at the horn tip, and the spectra consisted only of the broad continuum without Ar lines. This behavior may be related to the increased bubble–bubble interaction (Sects. 2.16 and 2.18). Further studies are required on this topic.

3.5   Ionization-Potential Lowering

As the density inside a bubble at the end of bubble collapse is in the same order of magnitude as the liquid density (condensed phase), plasma formed inside a bubble is under very high density compared to typical plasmas in the magnetic confinement

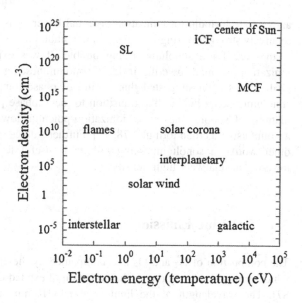

Fig. 3.14 Range of temperatures and densities of plasmas. 1 eV = 1.16 × 10⁴ K. SL, ICF, and MCF are sonoluminescence, inertial, and magnetic confinement fusion, respectively. The range for SL is similar to that of free electrons in solid metals

fusion (MCF) and in the universe except centers of fixed stars like the sun (Fig. 3.14) [64, 65]. A plasma is defined as a quasi-neutral gas of charged and neutral particles which exhibits collective behavior [66]. Any ionized gas cannot be called plasma because there is always a small degree of ionization in any gas. Quantitative criteria for plasmas are described in detail in Ref. [66]. The density of plasma inside a bubble is in the same order of magnitude or two orders of magnitude smaller than that in the inertial confinement fusion (ICF) using implosion of a spherical target containing fusion fuel by laser irradiation [67]. Some researchers of ICF are interested in acoustic cavitation because the densities of plasma inside a bubble are similar to those in ICF.

Furthermore, there have been some experimental reports that nuclear emissions were observed during acoustic cavitation in deuterated acetone [68–71]. However, there is an experimental report that nuclear emissions were not observed [72]. There is another experimental report that d + d → T + p reaction was accelerated in metal lithium by acoustic cavitation at ultrasonic frequency of ∼20 kHz with deuteron bombardment with its energy ranging from 30 to 70 keV [73]. Here, incident deuterons accumulated in liquid Li through beam bombardment and were regarded as additional targets. It was suggested that high temperatures inside bubbles caused the acceleration of d-d reaction. However, it is also possible that deuterons are accumulated inside the cavitation bubbles, and the local concentration of deuterons is increased. Further studies are required on this topic.

It has been known that ionization-potential lowering occurs in dense plasma [65, 74]. An extreme case of ionization-potential lowering is pressure ionization that ionization takes place by extremely high pressure [65]. The free electrons in metals in condensed phase (solid or liquid) originate in a kind of pressure ionization of atoms under high density [75]. For relatively weak lowering in ionization potential, an accurate formula for ionization-potential lowering is known [74]. However, for relatively strong lowering in ionization potential, little is known quantitatively. This is the case for a sonoluminescing bubble. It was experimentally reported that ionization-potential lowering inside a sonoluminescing bubble is by at least 75% [76]. It was even suggested that a kind of phase transition takes place inside a sonoluminescing bubble like transition to metal phase [76, 77]. In numerical simulations of sonoluminescence, ionization-potential lowering has been taken into account using a crude formula [78, 79]. Further studies are required on this topic. In other words, a sonoluminescing bubble is useful to study ionization-potential lowering at relatively high density.

3.6 OH-Line Emission

The combustion of H_2 and O_2 is accompanied by the emission of ultraviolet light, which is almost entirely due to the transition of excited OH to the ground state [80–87]. The wavelength of the light is about 310 nm, and the emission is called OH-line emission because it is not continuum in spectrum but a line. The OH-line emission has been experimentally observed in sonoluminescence, especially in MBSL [88–93]. It has also been experimentally observed from very dark SBSL [94]. Detailed mechanism of OH-line emission in SL is, however, still under debate as explained below.

The OH-line at about 310 nm is emitted when an electron in the first excited state (A state) of OH radical is de-excited to the ground state (X state). It is called OH (A–X) band because both A and X states have various vibrational and rotational states (Fig. 3.15) [95]. The ground state of an electron in a molecule is usually labeled X, and the excited states are labeled A, B, C, ... [63]. Vibrational states of a molecule are quantized as 0, 1, 2, 3, ... according to quantum mechanics as shown in Fig. 3.15. Rotational states of a molecule are also quantized as $J = 0, 1, 2, 3, ...$. As the rotational energy levels (10^{-4} to 10^{-2} eV) are much lower than those of the vibrational energy levels (0.1–1 eV), they are not shown in Fig. 3.15 [96, 97]. The electronic energy levels (X, A, B, C, ...) are in several eV corresponding to the energy of visible-to-ultraviolet light. The vibrational and rotational energies correspond to the energies of infrared light and microwave, respectively. The energy of electromagnetic wave (light) is given by $hc_{light}/\lambda_{light}$, where h is the Planck constant (6.6×10^{-34} Js), c_{light} is the speed of light (3.0×10^8 m/s), λ_{light} is the wavelength of light in meter, and 1 eV = 1.6×10^{-19} J.

The total angular momenta of electrons of 0, 1, 2, 3, ... are labeled Σ, Π, Δ, Φ, ... because the Greek letters correspond to S, P, D, F, ...(Fig. 3.15). The left superscript of a Greek letter shows $2S + 1$, where S is the total spin angular momentum. The symbol "+" in the right superscript means that the wave function of electrons does not change sign by reflection on the plane containing two nuclei of a molecule. ("−"means that the sign is changed.)

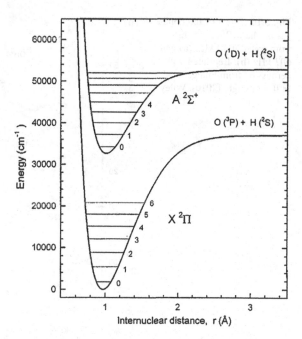

Fig. 3.15 Potentials for the A and X electronic states of OH. Reprinted with permission from Luque and Crosley [95]. Copyright (1998), AIP Publishing LLC

In MBSL, it has been experimentally reported that not only OH (A–X) band but also OH (C–A) band is observed in MBSL (Fig. 3.16) [89, 90, 93]. OH (C–A) band is in the range of 225–255 nm in wavelength and is emitted when an electron in the third excited state (C state) is de-excited to the first excited state (A state). The OH ($C\,^2\Sigma^+ - A\,^2\Sigma^+$) band has never been observed in normal combustion, while it has been observed in discharge in water vapor as well as γ-ray or electron irradiation of liquid water. Thus, it has been suggested that OH is excited to the third excited state by a collision with a high-energy electron created inside a bubble. It is an evidence of plasma formation inside a MBSL bubble.

The intensity of OH (C–A) band relative to that of OH (A–X) band depends on the noble gas species dissolved in water (Fig. 3.16) [89, 93]. From Xe bubbles, the OH (C–A) band is stronger than OH (A–X) band. From Ar bubbles, on the other hand, the OH (A–X) band is stronger than OH (C–A) band. The reason may be the larger number of high-energy electrons inside a Xe bubble due to lower ionization potential (ionization potentials are 12.1 and 15.8 eV for Xe and Ar, respectively) [98]. With regard to the difference of bubble temperature for different noble gases, some researchers reported that Xe bubbles are hotter than Ar bubbles [99, 100]. Some other researchers reported, however, that the temperatures inside Xe and Ar bubbles are nearly the same [98, 101]. This point should be studied in more detail in future. The relative intensity of OH (C–A) band also depends on ultrasonic frequency (Fig. 3.16) [89, 90, 93]. The detailed mechanism should be studied in future.

Fig. 3.16 Effect of the noble gas on the MBSL spectra from water at 20 kHz (**a**) and 607 kHz (**b**). Reprinted with permission from Pflieger et al. [89]. Copyright (2010), Wiley

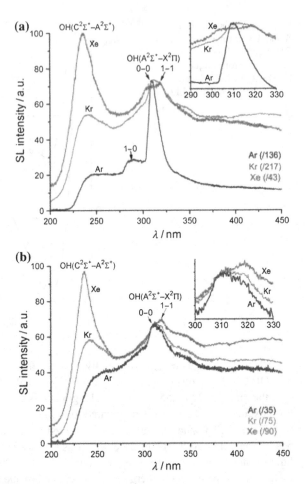

The vibrational population distribution of OH ($A\,^2\Sigma^+$) derived from the analysis of OH (A–X) band in MBSL spectra deviates significantly from the Boltzmann distribution (Fig. 3.17) [90, 93]. It means that the vibrational population of OH ($A\,^2\Sigma^+$) is in non-equilibrium inside MBSL bubbles. For equilibration, molecules and radicals should undergo enough number of collisions among them. The numbers of collisions necessary to provide equilibrium distributions from strongly perturbed thermodynamic states of an assembly of particles are in the order of 10, 10^3, and 10^5 for translational, rotational, vibrational motions, respectively [102]. For electronic excitation and dissociation of molecules and radicals, it is in the order of 10^7. For ionization, it is in the order of 10^9. Inside a sonoluminescing bubble, the number density of molecules and radicals is about 3×10^{28}/m^3 [103]. The average velocity of each molecule at 10^4 K is about 3×10^3 m/s. Thus, the frequency of collision for each molecule or radical is about 5×10^{13}/s = 0.6×10^{-18} m^2 (cross section of a molecule or radical) $\times 3 \times 10^3$ m/s (mean velocity) $\times 3 \times 10^{28}$/m^3

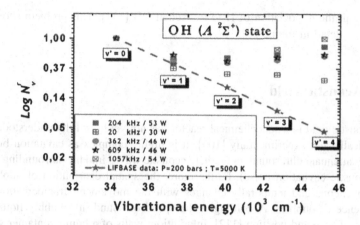

Fig. 3.17 Relative vibrational population distribution of the OH($A^2\Sigma^+$) state as a function of vibrational energy for different ultrasonic frequencies. The dashed line shows the equilibrium Boltzmann distribution. Reprinted with permission from Ndiaye et al. [90]. Copyright (2012), the American Chemical Society

(number density). The timescale for the temperature change inside a sonoluminescing bubble is in the order of 0.1 ns = 10^{-10} s [103]. The number of collisions of a molecule or a radical is about 5×10^3 during the time. It is enough for equilibration of translational and rotational motion of molecules and radicals. However, it is insufficient for the equilibration of vibrational population of molecules and radicals. Thus, vibrational population distribution deviates significantly from the equilibrium Boltzmann distribution inside a sonoluminescing bubble.

In the numerical simulations of chemical reactions inside a sonoluminescing bubble, not only translational and rotational motion but also vibrational motion of molecules and radicals is implicitly assumed to be in equilibrium [79, 103–108]. As vibrational population of molecules and radicals is strongly in non-equilibrium, however, numerical simulations of chemical reactions inside a sonoluminescing bubble should be performed in future taking into account the non-equilibrium effect. Note that non-equilibrium effect of chemical reactions has been taken into account by calculating both forward and backward reaction rates although the non-equilibrium population of vibrational states of molecules and radicals has not been taken into account in the calculations of the reaction rates [79, 103–108].

According to the experimentally derived population distribution of vibrational states of OH radicals, the population increases as the vibrational quantum number increases at relatively high quantum numbers as shown in Fig. 3.17 [90, 93]. The origin of such distribution is still unclear. One possibility is the excitation of vibrational states of OH by chemical reactions. It is widely known that molecular vibration is preferentially excited through some kinds of chemical reactions [96]. In this case, the electronic excitation of OH is also due to the chemical reactions. In other words, OH-line emission is chemiluminescence [78, 109, 110]. Another possibility is the excitation of H_2O by a collision with another molecule such as Ar, resulting in the dissociation of H_2O to excited OH [111]. This problem should be studied in detail in future.

3.7 Acoustic Field

The acoustic field in a sonochemical reactor has not yet been fully understood both theoretically and experimentally [109]. It is highly complex as cavitation bubbles strongly attenuate ultrasound and radiate acoustic waves into the surrounding liquid (the acoustic cavitation noise). Furthermore, the spatial distribution of bubbles is inhomogeneous and temporally changes with the movement, fragmentation, and coalescence of bubbles. Accordingly, the speed of sound in a bubbly liquid is a function of time and position [112]. In addition, walls of a liquid container vibrate due to the pressure oscillation of ultrasound. Vibrating walls radiate acoustic waves into the liquid, which also influences the acoustic field in the liquid [113].

An important problem is a role of large degassing bubbles in an acoustic field. When a liquid was irradiated with pulsed ultrasound, bubble size decreased and large degassing bubbles disappeared [114]. As a result, acoustic filed became more homogeneous in the liquid compared to that under irradiation of continuous ultrasound, which was confirmed by the observation of sonochemiluminescence (SCL). The detailed mechanism is, however, still unclear.

Another important problem is a role of liquid surface vibration in an acoustic field. There is an experimental report that liquid surface vibration occurred by strong ultrasound irradiation from the bottom of liquid container. When the amplitude of liquid surface vibration exceeded a quarter wavelength of ultrasound, SCL intensity strongly decreased [115]. This problem should be studied in more detail in future.

3.8 Effect of a Magnetic Field

In the experiment of single-bubble sonoluminescence (SBSL), it was experimentally reported by two research groups that SBSL intensity decreased as the magnetic flux density increased in the order of several tesla [116, 117]. Both the upper and lower bounds of acoustic amplitude for SBSL increased dramatically as the magnetic field increased. The dependence on magnetic field was different between liquid temperatures of 10 and 20 °C [116]. One of the groups reported that the maximum bubble radius decreased as the magnetic field increased [117]. The spectrum of SBSL gradually shifted to longer wavelength as the magnetic field increased [117]. By the addition of a surfactant (SDS) to liquid water, the effect of a magnetic field on SBSL almost vanished [117]. However, there is a report that the observed magnetic-field effect was not on the bubble itself but on the flask [118]. More detailed studies are required on this topic.

Theoretically, it has been suggested that moving water molecules interact with a magnetic field by the Lorentz force because water molecules possess a permanent electric dipole moment [119]. By the interaction, a part of the kinetic energy of liquid water around a pulsating bubble is dissipated as heat. The analytical calculations indicate that the effect of a magnetic field is similar to the effect of ambient pressure increase. It is also suggested that non-polar liquid such as dodecane exhibits no effect of a magnetic field.

3.9 Role of Oxygen Atoms

Numerical simulations of chemical reactions inside a bubble in water have shown that an appreciable amount of O atoms (radicals) are produced inside a bubble [107, 108]. However, the role of O atoms in sonochemical reactions in liquid water is still unclear [109]. The ground and the first excited states of O atom are O (^3P) and O (^1D), respectively, where P and D mean the total orbital angular momentum of 1 and 2, respectively, and the superscript means the multiplicity [109]. O atom at the first excited state immediately reacts with liquid water as follows:

$$O(^1D) + H_2O \rightarrow H_2O_2 \tag{3.7}$$

The rate constant for the reaction with H_2O vapor is $1.8 \pm 0.8 \times 10^{10}$ L/(mol s) [120]. Assuming this rate constant for liquid water, the lifetime of O (^1D) in liquid water is about 10^{-12} s = 1 ps. The diffusion length of O atom in this lifetime is only about 0.1 nm which is estimated by $2\sqrt{D_O \tau_O}$, where D_O is the diffusion coefficient of O atom in liquid water ($\sim 10^9$ m^2/s) and τ_O is the lifetime of O atom. Thus, O (^1D) is present only at the gas–liquid interface region of a bubble.

The ground state O (^3P) slowly reacts with molecules that have no unpaired electrons such as H_2O [109]. However, the reaction rate with H_2O is unknown at present. Further studies are required on a role of O (^3P) in sonochemical reactions in liquid water.

3.10 Extreme Conditions in a Dissolving Bubble

There are some experimental reports that OH radicals were detected from liquid water containing bulk nanobubbles (ultrafine bubbles) [121–123]. In order to study the possibility of radical formation from a dissolving bubble, numerical simulations of bubble dissolution have been performed taking into account the effect of bubble dynamics (the inertia of surrounding liquid) [124]. Surprisingly, the result indicated that temperature and pressure inside an air bubble increase up to 3000 K and 5 GPa, respectively, at the final stage of the bubble dissolution. At the final 2.3 ns before the complete dissolution of a bubble, the bubble content is only N_2 because the solubility of N_2 is the lowest among the gases (N_2, O_2, Ar) and the pressure inside a bubble is many orders of magnitude higher than the saturated vapor pressure of H_2O. The liquid temperature at the bubble wall increases up to 85 °C at the final moment of the bubble dissolution. This liquid temperature is insufficient for the thermal dissociation of H_2O molecules. Inside a bubble, the probability of the dissociation of N_2 molecules is only on the order of 10^{-15} at 3000 K at the final moment partly due to the very short duration of high temperature. At present, however, there is no experimental confirmation of the result. Further studies are required on this topic including numerical simulations for other gases such as a pure O_2 bubble and an O_3 bubble.

3.11   Concluding Remarks (Modeling Complex Phenomena)

Complex phenomena such as acoustic cavitation are not always suitable for the first principle calculations such as computational fluid-dynamics simulations (Navier–Stokes equation), molecular dynamics simulations, FEM (finite element method) calculations. In many cases, it is better to make a suitable theoretical model taking into account all the important effects. The accuracy of numerical simulations based on such a theoretical model may be generally worse than that of first principle calculations. However, such a theoretical model is more suitable to understand qualitative meaning of the phenomena, mechanism of some experimental results, as well as to obtain theoretical predictions by simulating under various conditions because such simulations are computationally more economical compared to the first principle calculations, and the important factors are more easily traced. Not only experiments but also computer simulations are processes of knowledge creation in the present age of computer simulation [125].

References

1. Fisher JC (1948) The fracture of liquids. J Appl Phys 19:1062–1067. doi:10.1063/1.1698012
2. Temperley HNV (1947) The behaviour of water under hydrostatic tension: III. Proc Phys Soc London 59:199–208. doi:10.1088/0959-5309/59/2/304
3. Yasui K (2015) Dynamics of acoustic bubbles. In: Grieser F, Choi PK, Enomoto N, Harada H, Okitsu K, Yasui K (eds) Sonochemistry and the acoustic bubble. Elsevier, Amsterdam
4. Galloway WJ (1954) An experimental study of acoustically induced cavitation in liquids. J Acoust Soc Am 26:849–857. doi:10.1121/1.1907428
5. Epstein PS, Plesset MS (1950) On the stability of gas bubbles in liquid-gas solutions. J Chem Phys 18:1505–1509. doi:10.1063/1.1747520

6. Alheshibri M, Qian J, Jehannin M, Craig VSJ (2016) A history of nanobubbles. Langmuir 32:11086–11100. doi:10.1021/acs.langmuir.6b02489
7. Kobayashi H, Maeda S, Kashiwa M, Fujita T (2014) Measurement and identification of ultrafine bubbles by resonant mass measurement method. Proc SPIE 9232:92320S. doi:10.1117/12.2064811
8. Kobayashi H, Maeda S, Kashiwa M, Fujita T (2014) Measurement of ultrafine bubbles using different types of particle size measuring instruments. Proc SPIE 9232:92320U. doi:10.1117/12.2064638
9. Tuziuti T, Yasui K, Kanematsu W (2017) Influence of increase in static pressure on bulk nanobubbles. Ultrason Sonochem 38:347–350. doi:10.1016/j.ultsonch.2017.03.036
10. Tsuge H (ed) (2014) Micro- and nanobubbles. Pan Stanford, Singapore
11. Yount DE (1979) Skins of varying permeability: a stabilization mechanism for gas cavitation nuclei. J Acoust Soc Am 65:1429–1439. doi:10.1121/1.382930
12. Fox FE, Herzfeld KF (1954) Gas bubbles with organic skin as cavitation nuclei. J Acoust Soc Am 26:984–989. doi:10.1121/1.1907466
13. Yasui K (2016) Mechanism for stability of ultrafine bubbles. Jpn J Multiph Flow 30:19–26 (in Japanese)
14. Takahashi M (2005) ζ potential of microbubbles in aqueous solutions: electrical properties of the gas-water interface. J Phys Chem B 109:21858–21864. doi:10.1021/jp0445270
15. Oh SH, Han JG, Kim JM (2015) Long-term stability of hydrogen nanobubble fuel. Fuel 158:399–404. doi:10.1016/j.fuel.2015.05.072
16. Cho SH, Kim JY, Chun JH, Kim JD (2005) Ultrasonic formation of nanobubbles and their zeta-potentials in aqueous electrolyte and surfactant solutions. Colloids Surf, A 269:28–34. doi:10.1016/j.colsurfa.2005.06.063
17. Kim JY, Song MG, Kim JD (2000) Zeta potential of nanobubbles generated by ultrasonication in aqueous alkyl polyglycoside solutions. J Colloid Interface Sci 223:285–291. doi:10.1006/jcis.1999.6663
18. Steitz R, Gutberlet T, Hauss T, Klosgen B, Krastev R, Schemmel S, Simonsen AC, Findenegg GH (2003) Nanobubbles and their precursor layer at the interface of water against a hydrophobic substrate. Langmuir 19:2409–2418. doi:10.1021/la026731p
19. Mezger M, Schoder S, Reichert H, Schroder H, Okasinski J, Honkimaki V, Ralston J, Bilgram J, Roth R, Dosch H (2008) J Chem Phys 128:244705. doi:10.1063/1.2931574
20. Lu YH, Yang CW, Hwang IS (2012) Molecular layer of gaslike domains at a hydrophobic-water interface observed by frequency-modulation atomic force microscopy. Langmuir 28:12691–12695. doi:10.1021/la301671a
21. Peng H, Birkett GR, Nguyen AV (2013) Origin of interfacial nanoscopic gaseous domains and formation of dense gas layer at hydrophobic solid-water interface. Langmuir 29:15266–15274. doi:10.1021/la403187p
22. Azadi M, Nguyen AV, Yakubov GE (2015) Attractive forces between hydrophobic solid surfaces measured by AFM on the first approach in salt solutions and in the presence of dissolved gases. Langmuir 31:1941–1949. doi:10.1021/la504001z
23. Peng H, Hampton MA, Nguyen AV (2013) Nanobubbles do not sit alone at the solid-liquid interface. Langmuir 29:6123–6130. doi:10.1021/la305138v
24. Yasui K, Tuziuti T, Kanematsu W, Kato K (2016) Dynamic equilibrium model for a bulk nanobubble and a microbubble partly covered with hydrophobic material. Langmuir 32:11101–11110. doi:10.1021/acs.langmuir.5b04703
25. Yasui K, Tuziuti T, Kanematsu W, Kato K (2015) Advanced dynamic-equilibrium model for a nanobubble and a micropancake on a hydrophobic or hydrophilic surface. Phys Rev E 91:033008. doi:10.1103/PhysRevE.91.033008
26. Petsev ND, Shell MS, Leal LG (2013) Dynamic equilibrium explanation for nanobubbles' unusual temperature and saturation dependence. Phys Rev E 88:010402(R). doi:10.1103/PhysRevE.88.010402
27. Israelachvili JN (2011) Intermolecular and surface forces, 3rd edn. Elsevier, Amsterdam

28. Lee J, Tuziuti T, Yasui K, Kentish S, Grieser F, Ashokkumar M, Iida Y (2007) Influence of surface-active solutes on the coalescence, clustering, and fragmentation of acoustic bubbles confined in a microspace. J Phys Chem C 111:19015–19023. doi:10.1021/jp075431j

29. Chan CU, Ohl CD (2012) Total-internal-reflection-fluorescence microscopy for the study of nanobubble dynamics. Phys Rev Lett 109:174501. doi:10.1103/PhysRevLett.109.174501

30. Dietrich E, Zandvliet HJW, Lohse D, Seddon JRT (2013) Particle tracking around surface nanobubble. J Phys: Condens Matter 25:184009. doi:10.1088/0953-8984/25/18/184009

31. Lohse D, Zhang X (2015) Surface nanobubbles and nanodroplets. Rev Mod Phys 87:981–1035. doi:10.1103/RevModPhys.87.981

32. Brenner MP, Lohse D (2008) Dynamic equilibrium mechanism for surface nanobubble stabilization. Phys Rev Lett 101:214505. doi:10.1103/PhysRevLett.101.214505

33. Modell M, Reid RC (1974) Thermodynamics and its applications. Prentice Hall, Englewood Cliffs, NJ

34. Kondepudi D, Prigogine I (1998) Modern thermodynamics. John Wiley & Sons, Chichester, UK

35. Lohse D, Zhang X (2015) Pinning and gas oversaturation imply stable single surface nanobubble. Phys Rev E 91:031003 (R). doi:10.1103/PhysRevE.91.031003

36. Ouerhani T, Pflieger R, Messaoud WB, Nikitenko SI (2015) Spectroscopy of sonoluminescence and sonochemistry in water saturated with N_2–Ar mixtures. J Phys Chem B 119:15885–15891. doi:10.1021/acs.jpcb.5b10221

37. Hart EJ, Fischer CH, Henglein A (1986) Isotopic exchange in the sonolysis of aqueous solutions containing $^{14,14}N_2$ and $^{15,15}N_2$. J Phys Chem 90:5989–5991. doi:10.1021/j100280a104

38. Supeno Kruus P (2002) Fixation of nitrogen with cavitation. Ultrason Sonochem 9:53–59. doi:10.1016/S1350-4177(01)60070-0

39. Sokol'skaya AV (1978) Synthesis of glycine from formaldehyde and molecular nitrogen in aqueous solutions in an ultrasonic field. J Gen Chem USSR 48:1289–1292

40. Hickling R, Plesset MS (1964) Collapse and rebound of a spherical bubble in water. Phys Fluids 7:7–14. doi:10.1063/1.1711058

41. Hickling R (1994) Transient, high-pressure solidification associated with cavitation in water. Phys Rev Lett 73:2853–2856. doi:10.1103/PhysRevLett.73.2853

42. Hickling R (1965) Nucleation of freezing by cavity collapse and its relation to cavitation damage. Nature (London) 206:915–917. doi:10.1038/206915a0

43. Sukovich JR, Anderson PA, Sampathkumar A, Gaitan DF, Pishchalnikov YA, Holt RG (2017) Outcomes of the collapse of a large bubble in water at high ambient pressures. Phys Rev E 95:043101. doi:10.1103/PhysRevE.95.043101

44. Chow R, Blindt R, Chivers R, Povey M (2003) The sonocrystallisation of ice in sucrose solutions: primary and secondary nucleation. Ultrasonics 41:595–604. doi:10.1016/j.ultras.2003.08.001

45. Chow R, Mettin R, Lindinger B, Kurz T, Lauterborn W (2003) The importance of acoustic cavitation in the sonocrystallisation of ice—high speed observation of a single acoustic bubble. In: 2003 IEEE Ultrasonics Symposium Proceeding, vol 2 pp 1447–1450. doi:10.1109/ULTSYM.2003.1293177

46. Chow R, Blindt R, Kamp A, Grocutt P, Chivers R (2004) The microscopic visualization of the sonocrystallisation of ice using a novel ultrasonic cold stage. Ultrason Sonochem 11:245–250. doi:10.1016/j.ultsonch.2004.01.018

47. Chow R, Blindt R, Chivers R, Povey M (2005) A study on the primary and secondary nucleation of ice by power ultrasound. Ultrasonics 43:227–230. doi:10.1016/j.ultras.2004.06.006

48. Lindinger B, Mettin R, Chow R, Lauterborn W (2007) Ice crystallization induced by optical breakdown. Phys Rev Lett 99:045701. doi:10.1103/PhysREvLett.99.045701

49. Wohlgemuth K, Ruether F, Schembecker G (2010) Sonocrystallization and crystallization with gassing of adipic acid. Chem Eng Sci 65:1016–1027. doi:10.1016/j.ces.2009.09.055

50. Soare A, Dijkink R, Pascual MR, Sun C, Cains PW, Lohse D, Stankiewicz AI, Kramer HJM (2011) Crystal nucleation by laser-induced cavitation. Cryst Growth Des 11:2311–2316. doi:10.1021/cg2000014

51. Jamshidi R, Rossi D, Saffari N, Gavriilidis A, Mazzei L (2016) Investigation of the effect of ultrasound parameters on continuous sonocrystallization in a millifluidic device. Cryst Growth Des 16:4607–4619. doi:10.1021/acs.cgd.6b00696

52. Castillo-Peinado LS, Dolores M, Castro L (2016) The role of ultrasound in pharmaceutical production: sonocrystallization. J Pharm Pharmacol 68:1249–1267. doi:10.1111/jphp.12614

53. Bhangu SK, Ashokkumar M, Lee J (2016) Ultrasound assisted crystallization of paracetamol: crystal size distribution and polymorph control. Cryst Growth Des 16:1934–1941. doi:10.1021/acs.cgd.5b01470

54. Wohlgemuth K, Kordylla A, Ruether F, Schembecker G (2009) Experimental study of the effect of bubbles on nucleation during batch cooling crystallization. Chem Eng Sci 64:4155–4163. doi:10.1016/j.ces.2009.06.041

55. Kordylla A, Krawczyk T, Tumakaka F, Schembecker G (2009) Modeling ultrasound-induced nucleation during cooling crystallization. Chem Eng Sci 64:1635–1642. doi:10.1016/j.ces.2008.12.030

56. Yasui K, Tuziuti T, Kato K (2011) Numerical simulations of sonochemical production of $BaTiO_3$ nanoparticles. Ultrason Sonochem 18:1211–1217. doi:10.1016/j.ultsonch.2011.03.006

57. Moss WC, Clarke DB, Young DA (1997) Calculated pulse widths and spectra of a single sonoluminescing bubble. Science 276:1398–1401. doi:10.1126/science.276.5317.1398

58. Moss WC, Young DA, Harte JA, Levatin JL, Rozsnyai BF, Zimmerman GB, Zimmerman IH (1999) Computed optical emissions from a sonoluminescing bubble. Phys Rev E 59:2986–2992. doi:10.1103/PhysRevE.59.2986

59. Burnett PDS, Chambers DM, Heading D, Machacek A, Moss WC, Rose S, Schnittker M, Lee RW, Young P, Wark JS (2001) Modeling a sonoluminescing bubble as a plasma. J Quant Spectro Radiat Transfer 71:215–223. doi:10.1016/S0022-4073(01)00069-3

60. Flannigan DJ, Suslick KS (2005) Plasma formation and temperature measurement during single-bubble cavitation. Nature (London) 434:52–55. doi:10.1038/nature03361

61. Eddingsaas NC, Suslick KS (2007) Evidence for a plasma core during multibubble sonoluminescence in sulfuric acid. J Am Chem Soc 129:3838–3839. doi:10.1021/ja070192z

62. Fridman A (2008) Plasma chemistry. Cambridge Univ Press, Cambridge

63. Atkins PW, Friedman RS (1997) Molecular quantum mechanics. Oxford Univ Press, Oxford

64. Peratt AL (2015) Physics of the plasma universe, 2nd edn. Springer, New York

65. Salzmann D (1998) Atomic physics in hot plasmas. Oxford Univ Press, Oxford

66. Chen FF (2016) Introduction to plasma physics and controlled fusion, 3rd edn. Springer, Cham, Switzerland

67. Craxton RS, Anderson KS, Boehly TR, Goncharov VN, Harding DR, Knauer JP, McCrory RL, McKenty PW, Meyerhofer DD, Myatt JF, Schmitt AJ, Sethian JD, Short RW, Skupsky S, Theobald W, Kruer WL, Tanaka K, Betti R, Collins TJB, Delettrez JA, Hu SX, Marozas JA, Maximov AV, Michel DT, Radha PB, Regan SP, Sangster TC, Seka W, Solodov AA, Soures JM, Stoeckl JM, Zuegel JD (2015) Direct-drive inertial confinement fusion: a review. Phys Plasmas 22:110501. doi:10.1063/1.4934714

68. Taleyarkhan RP, West CD, Cho JS, Lahey RT Jr, Nigmatulin RI, Block RC (2002) Evidence for nuclear emissions during acoustic cavitation. Science 295:1868–1873. doi:10.1126/science.1067589

69. Taleyarkhan RP, Cho JS, West CD, Lahey RT Jr, Nigmatulin RI, Block RC (2004) Additional evidence of nuclear emissions during acoustic cavitation. Phys Rev E 69:036109. doi:10.1103/PhysRevE.69.036109

70. Taleyarkhan RP, West CD, Lahey RT Jr, Nigmatulin RI, Block RC, Xu Y (2006) Nuclear emissions during self-nucleated acoustic cavitation. Phys Rev Lett 96:034301. doi:10.1103/PhysRevLett.96.034301

71. Xu Y, Butt A (2005) Confirmatory experiments for nuclear emissions during acoustic cavitation. Nuclear Eng Design 235:1317–1324. doi:10.1016/j.nucengdes.2005.02.021

72. Camara CG, Hopkins SD, Suslick KS, Putterman SJ (2007) Upper bound for neutron emission from sonoluminescing bubbles in deuterated acetone. Phys Rev Lett 98:064301. doi:10.1103/PhysRevLett.98.064301

73. Toriyabe Y, Yoshida E, Kasagi J, Fukuhara M (2012) Acceleration of the d + d reaction in metal lithium acoustic cavitation with deuteron bombardment from 30 to 70 keV. Phys Rev C 85:054620. doi:10.1103/PhysRevC.85.054620

74. Zel'dovich YB, Raizer YP (2002) Physics of shock waves and high-temperature hydrodynamic phenomena. Dover, Mineola, New York

75. Ashcroft NW, Mermin ND (1976) Solid state physics. Sounders College, Philadelphia

76. Kappus B, Bataller A, Putterman SJ (2013) Energy balance for a sonoluminescence bubble yields a measure of ionization potential lowering. Phys Rev Lett 111:234301. doi:10.1103/PhysRevLett.111.234301

77. Kappus B, Khalid S, Chakravarty A, Putterman S (2011) Phase transition to an opaque plasma in a sonoluminescing bubble. Phys Rev Lett 106:234302. doi:10.1103/PhysREvLett.106.234302

78. Yasui K (2001) Effect of liquid temperature on sonoluminescence. Phys Rev E 64:016310. doi:10.1103/PhysRevE.64.016310

79. Yasui K, Tuziuti T, Lee J, Kozuka T, Towata A, Iida Y (2008) The range of ambient radius for an active bubble in sonoluminescence and sonochemical reactions. J Chem Phys 128:184705. doi:10.1063/1.2919119

80. Gaydon AG, Wolfhard HG (1951) Predissociation in the spectrum of OH; the vibrational and rotational intensity distribution in flames. Proc Roy Soc A 208:63–75. doi:10.1098/rspa.1951.0144

81. Broida HP, Shuler KE (1952) Kinetics of OH radicals from flame emission spectra. IV. A study of the hydrogen-oxygen flame. J Chem Phys 20:168–174. doi:10.1063/1.1700163

82. Charton M, Gaydon AG (1958) Excitation of spectra of OH in hydrogen flames and its relation to excess concentrations of free atoms. Proc Roy Soc A 245:84–92. doi:10.1098/rspa.1958.0068

83. Kaskan WE (1959) Abnormal excitation of OH in $H_2/O_2/N_2$ flames. J Chem Phys 31:944–956. doi:10.1063/1.1730556

84. Belles FE, Lauver MR (1964) Origin of OH chemiluminescence during the induction period of the H_2–O_2 reaction behind shock waves. J Chem Phys 40:415–422. doi:10.1063/1.1725129

85. Gutman D, Lutz RW, Jacobs NF, Hardwidge EA, Schott GL (1968) Shock-tube study of OH chemiluminescence in the hydrogen-oxygen reaction. J Chem Phys 48:5689–5694. doi:10.1063/1.1668656

86. Davis MG, McGregor WK, Mason AA (1974) OH chemiluminescent radiation from lean hydrogen-oxygen flames. J Chem Phys 61:1352–1356. doi:10.1063/1.1682059

87. Hidaka Y, Takahashi S, Kawano H, Suga M, Gardiner WC Jr (1982) Shock-tube measurement of the rate constant for excited $OH(A^2\Sigma^+)$ formation in the hydrogen-oxygen reaction. J Phys Chem 86:1429–1433. doi:10.1021/j100397a043

88. Matula TJ, Roy RA, Mourad PD, McNamara WB III, Suslick KS (1995) Comparison of multibubble and single-bubble sonoluminescence spectra. Phys Rev Lett 75:2602–2605. doi:10.1103/PhysRevLett.75.2602

89. Pflieger R, Brau HP, Nikitenko SI (2010) Sonoluminescence from $OH(C^2\Sigma^+)$ and $OH(A^2\Sigma^+)$ radicals in water: evidence for plasma formation during multibubble cavitation. Chem Eur J 16:11801–11803. doi:10.1002/chem.201002170

90. Ndiaye AA, Pflieger R, Siboulet B, Molina J, Dufreche JF, Nikitenko SI (2012) Nonequilibrium vibrational excitation of OH radicals generated during multibubble cavitation in water. J Phys Chem A 116:4860–4867. doi:10.1021/jp301989b

91. Flannigan DJ, Suslick KS (2013) Non-Boltzmann population distributions during single-bubble sonoluminescence. J Phys Chem B 117:15886–15893. doi:10.1021/jp409222x

92. Pflieger R, Ndiaye AA, Chave T, Nikitenko SI (2015) Influence of ultrasonic frequency on swan band sonoluminescence and sonochemical activity in aqueous tert-butyl alcohol solutions. J Phys Chem B 119:284–290. doi:10.1021/jp509898p

93. Nikitenko SI, Pflieger R (2017) Toward a new paradigm for sonochemistry: short review on nonequilibrium plasma observations by means of MBSL spectroscopy in aqueous solutions. Ultrason Sonochem 35:623–630. doi:10.1016/j.ultsonch.2016.02.003

94. Young JB, Nelson JA, Kang W (2001) Line emission in single-bubble sonoluminescence. Phys Rev Lett 86:2673–2676. doi:10.1103/PhysRevLett.86.2673

95. Luque J, Crosley DR (1998) Transition probabilities in the $A^2\Sigma^+\!-\!X^2\Pi_i$ electronic system of OH. J Chem Phys 109:439–448. doi:10.1063/1.476582

96. McQuarrie DA, Simon JD (1997) Physical chemistry: a molecular approach. University Science, Sausalito

97. Atkins P, Paula J (2014) Atkins' physical chemistry, 10th edn. Oxford University Press, Oxford

98. Yasui K (2002) Segregation of vapor and gas in a sonoluminescing bubble. Ultrasonics 40:643–647. doi:10.1016/S0041-624X(02)00190-7

99. McNamara WB III, Didenko YT, Suslick KS (1999) Sonoluminescence temperatures during multi-bubble cavitation. Nature (London) 401:772–775

100. Didenko YT, McNamara WB III, Suslick KS (2000) Effect of noble gases on sonoluminescence temperatures during multibubble cavitation. Phys Rev Lett 84:777–780. doi:10.1103/PhysRevLett.84.777

101. Okitsu K, Suzuki T, Takenaka N, Bandow H, Nishimura R, Maeda Y (2006) Acoustic multibubble cavitation in water: a new aspect of the effect of a rare gas atmosphere on bubble temperature and its relevance to sonochemsitry. J Phys Chem B 110:20081–20084. doi:10.1021/jp064598u

102. Mavrodineanu R, Boiteux H (1965) Flame spectroscopy. John Wiley & Sons, New York

103. Yasui K (1997) Alternative model of single-bubble sonoluminescence. Phys Rev E 56:6750–6760. doi:10.1103/PhysRevE.56.6750

104. Storey BD, Szeri AJ (2000) Water vapour, sonoluminescence and sonochemistry. Proc R Soc Lond A 456:1685–1709. doi:10.1098/rspa.2000.0582

105. Storey BD, Szeri AJ (2001) A reduced model of cavitation physics for use in sonochemistry. Proc R Soc Lond A 457:1685–1700. doi:10.1098/rspa.2001.0784

106. Toegel R, Lohse D (2003) Phase diagrams for sonoluminescing bubbles: a comparison between experiment and theory. J Chem Phys 118:1863–1875. doi:10.1063/1.1531610

107. Yasui K, Tuziuti T, Sivakumar M, Iida Y (2005) Theoretical study of single-bubble sonochemistry. J Chem Phys 122:224706. doi:10.1063/1.1925607

108. Yasui K, Tuziuti T, Kozuka T, Towata A, Iida Y (2007) Relationship between the bubble temperature and main oxidant created inside an air bubble under ultrasound. J Chem Phys 127:154502. doi:10.1063/1.2790420

109. Yasui K (2016) Unsolved problems in acoustic cavitation. In: Ashokkumar M, Cavalieri F, Chemat F, Okitsu K, Sambandam A, Yasui K, Zisu B (eds) Handbook of ultrasonics and sonochemistry. Springer, Singapore

110. Yasui K (2001) Temperature in multibubble sonoluminescence. J Chem Phys 115:2893–2896. doi:10.1063/1.1395056

111. Didenko YT, Pugach SP (1994) Spectra of water sonoluminescence. J Phys Chem 98:9742–9749. doi:10.1021/j100090a006

112. Dahnke S, Keil F (1998) Modeling of sound fields in liquids with a nonhomogeneous distribution of cavitation bubbles as a basis for the design of sonochemical reactors. Chem Eng Technol 21:873–877. doi:10.1002/(SICI)1521-4125(199811)21:11<873:AID-CEAT873>3.0.CO;2-V

113. Yasui K, Kozuka T, Tuziuti T, Towata A, Iida Y, King J, Macey P (2007) FEM calculation of an acoustic field in a sonochemical reactor. Ultrason Sonochem 14:605–614. doi:10.1016/j.ultsonch.2006.09.010

114. Tuziuti T, Yasui K, Lee J, Kozuka T, Towata A, Iida Y (2008) Mechanism of enhancement of sonochemical-reaction efficiency by pulsed ultrasound. J Phys Chem A 112:4875–4878. doi:10.1021/jp802640x

115. Tuziuti T, Yasui K, Kozuka T, Towata A (2010) Influence of liquid-surface vibration on sonochemiluminescence intensity. J Phys Chem A 114:7321–7325. doi:10.1021/jp101638c

116. Young JB, Schmiedel T, Kang W (1996) Sonoluminescence in high magnetic fields. Phys Rev Lett 77:4816–4819. doi:10.1103/PhysRevLett.77.4816

117. Abe Y, Nakabayashi S (2002) SBSL dynamics under high magnetic field. In: Proceeding International Symposium on Innovative Materials Processing by Controlling Chemical Reaction Field (IMP2002), The Society of Non-Traditional Technology, Tokyo, pp 17–20

118. Brenner MP, Hilgenfeldt S, Lohse D (2002) Single-bubble sonoluminescence. Rev Mod Phys 74:425–484. doi:10.1103/RevModPhys.74.425

119. Yasui K (1999) Effect of a magnetic field on sonoluminescence. Phys Rev E 60:1759–1761. doi:10.1103/PhysRevE.60.1759

120. Biedenkapp D, Hartshorn LG, Bair EJ (1970) The O (^{1}D) + H_2O reaction. Chem Phys Lett 5:379–380. doi:10.1016/0009-2614(70)85172-7

121. Takahashi M, Chiba K, Li P (2007) Free-radical generation from collapsing microbubbles in the absence of a dynamic stimulus. J Phys Chem B 111:1343–1347. doi:10.1021/jp0669254

122. Liu S, Oshita S, Makino Y, Wang QH, Kawagoe Y, Uchida T (2016) Oxidative capacity of nanobubbles and its effect on seed germination. ACS Sustain Chem Eng 4:1347–1353. doi:10.1021/acssuschemeng.5b01368

123. Liu S, Oshita S, Kawabata S, Makino Y, Yoshimoto T (2016) Identification of ROS produced by nanobubbles and their positive and negative effects on vegetable seed germination. Langmuir 32:11295–11302. doi:10.1021/acs.langmuir.6b01621

124. Yasui K, Tuziuti T, Kanematsu W (2016) Extreme conditions in a dissolving air nanobubble. Phys Rev E 94:013106. doi:10.1103/PhysRevE.94.013106

125. Winsberg E (2010) Science in the age of computer simulation. University of Chicago Press, Chicago

Printed in the United States
By Bookmasters